ISBN 978-0-484-61014-8
PIBN 10098104

GEOLOGY

AND

INDUSTRIAL RESOURCES

OF

CALIFORNIA:

BY

PHILIP T. TYSON.

TO WHICH IS ADDED,

THE OFFICIAL REPORTS OF GENLS. PERSIFER F. SMITH AND B. RILEY—
INCLUDING THE REPORTS OF LIEUTS. TALBOT, ORD, DERBY AND
WILLIAMSON, OF THEIR EXPLORATIONS IN CALIFORNIA AND
OREGON; AND ALSO OF THEIR EXAMINATIONS OF
ROUTES FOR RAIL ROAD COMMUNICATION
EASTWARD FROM THOSE COUNTRIES.

BALTIMORE:
PUBLISHED BY WM. MINIFIE & CO.
114 Baltimore Street.

1851.

213341

ADVERTISEMENT.

THIS work mainly consists of the Memoir upon the Geology and Resources of California, reported to the U. S. Topographical Bureau by Mr. Tyson, and the official communications of Gens. P. F. Smith and B. Riley, accompanied by the reports of explorations in various parts of California and Oregon, by Lieut's Talbot, Ord, Derby, Capt. Warner and Lieut. Williamson.

Whilst the issue of these valuable reports has been long delayed, the style in which the work has been executed, and the materials employed, are not such as their importance entitles them to.

But, notwithstanding these mechanical imperfections, the publishers have great pleasure in presenting to the reader, a work of so much intrinsic value—and one which, it may be confidently believed, contains an amount of accurate and reliable information in regard to California and Oregon, more important than has hitherto been published. That it is of the highest moment to put the public in possession of authentic information in reference to California especially, will be generally admitted. We have ample reason for believing that the present work will materially aid in this respect; not only from the known ability of the authors of the several parts that compose it, but from the character of the reports themselves, which furnish ample testimony that the parties are so much accustomed to examine into the actual facts that come under their notice, that they are little liable to fall into errors—too frequent with many who have written of California within the past two years.

This edition consists of the copies presented to Mr. Tyson by the Senate of the United States, which have been put into our hands by that gentleman; and he has also enabled us to add materially to the value of the work, by important additional matter.

This *addition* is, in fact, *supplementary*; but the arrangement of the several Reports is such, that it could not immediately succeed the Memoir, without interrupting the numerical order of the pages; it is consequently placed as an *Introduction*.

The utility of this edition is not only increased by containing this supplementary matter, but by the opportunity it has afforded the author of the Memoir, to correct sundry errors that have crept into the portion of the work printed by order of the Senate. Some of these were altogether too important to be passed over. The reader is particularly requested to notice them in the errata, on page 2; several of them are such as to change altogether the meaning of sentences.

INDEX.

2

CONTENTS.

INTRODUCTION.

THE Report which forms the first portion of this work, was written more than nine months since, and, as it was expected to have been published without delay, its preparation was hastened in a greater degree than would otherwise have been desirable. Whether the public has gained or lost by the procrastination, the reader may decide; but the result appears to indicate that the machinery of government, (sometimes at least,) furnishes a tardy means of giving publicity to one's lucubrations, and therefore not always the best, unless the period at which a work shall appear, be a matter of indifference.

The object of the Report was to present a concise account of the resources of a region towards which public attention was strongly turned, and which, it was abundantly evident, was viewed through such a mixture of fiction and truth, as to present itself in a position extremely false.

Besides giving accounts of the metalliferous, agricultural and other industrial resources of California, suited to the general reader, it also contains such notices of the prominent geological features of the country traversed, as will, with the aid of the map and sections, enable the scientific inquirer to compare its structure with that of other regions. The facts were detailed as they appeared to be, and such observations made thereon, as were deemed necessary to illustrate them, and furnish information useful to the public.

In the letter to Col. Abert, (on page 2,) reference was made to the specimens collected by me in California, and which, to my great disappointment, did not arrive in the ship Andalusia,—whose Captain engaged to bring them to Baltimore. Whether the efforts he has made since his arrival to repair the mishap, will be successful or not, remains to be seen. Although they were mostly objects of scientific interest, yet there were among them vegetable, as well as mineral substances, supposed to possess an industrial importance.

3

In the same letter I acknowledged my indebtedness to officers of the army and navy, for facilities furnished by them whilst pursuing my investigations in California, but I omitted to acknowledge the kindness of Gov. Marcy, whilst he was Secretary of War, for a Circular, to officers commanding posts in California, requesting that facilities might be afforded whilst these examinations were in progress.

I found every where the officers, both of the army and navy, so ready to facilitate such investigations, that there was no necessity for exhibiting the Circular, and it was only shown to General P. F. Smith.

In reference to the subject that seemed most to interest the public hitherto,—that of the " gold diggings " and washings,—I was irresistibly led to the conclusion, that a very small proportion indeed of those who occupied themselves in collecting the metal from *the earth* were adequately rewarded, whilst the great body of them have done little, if any, more than to support themselves. And yet, the severity of the labor, the privations and incidental personal exposure, are unequalled by any pursuit practised in our country. And as a necessary consequence, disease and death prevailed so extensively, as to bring distress and want upon many a family at home, whose members had been induced to hasten to El Dorado, under the hope of soon returning with wealth in abundance. The chances of getting rich by these means, did not appear better than those of the lottery adventurer, who, in general, loses his money without impairing his health.

It was also shown, that it was to the exposure and privations that the sickness of last year was mainly owing, and not to the climate of the country, which, it may be said, will compare favorably, in reference to health, with that of any other country of similar extent within the same latitudes.

It is true that intermittent and other fevers prevail in summer and autumn on some portions of Sacramento and San Joaquin rivers, but the area of such infected districts is not extensive.

Some of the resources other than gold, were enlarged upon for the purpose of showing their importance, and in order to aid in turning attention to pursuits, in which industry was more likely to meet with certain reward, and in greater amount in the long run, than is the average lot of the *direct* gold-seeker.

When I left California in October, 1849, the mania for gold digging was rapidly declining, and the public here, as well as there, were beginning to learn something of its realities. The returning steamers were crowded, and many sailing vessels were also employed in transporting passengers home *via* Panama, whilst so many of those who remained were engaging, or endeavoring to engage in agricultural and other pursuits, as to indicate the near approach of the period when the various branches of industry, suited to the country, would draw off from the gold region a due proportion of labor.

There did not appear to be any probability of further causes for excitement likely to induce, during the present year, a greater immigration than would be advantageous to the parties or to the country.

But that there was still danger to be apprehended of a renewal of the excitement, from the indiscretion of writers or some other cause, was intimated by the observations on pages 4 and 33, which were intended as a caution against giving too easy credence to all accounts of an extravagant character in reference to the minerals of California, and was induced by the singular proneness, (believed to exist in the minds of most persons who have not systematically and practically studied such subjects,) to mistake them altogether; and to form the most erroneous opinions in regard to the quantity, quality, and even the kinds of minerals themselves. Any one, who has long devoted himself to the sciences relating to minerals, could bring forward a multitude of facts in proof of what is here stated.

Nothwithstanding the danger of further causes for excitements in the public mind, it was a matter of surprise that they came in the shape they did.

Just as the writing of the Memoir was being brought to a close, a copy of the Pacific News reached this country, containing an account of new wonders said to have been discovered. We were informed that "*gold bearing quartz was said to be found in inexhaustible masses or quarries through the whole mountainous region, which form the western slope of the Sierra Nevada,*" and that these "*quartz mountain quarries*" contained an average amount of gold, equal in value to $2 50 to $3 to the pound of rock!

This news was apparently of a startling character; but after making such allowances for the excitement of the writer or his informants, as were suggested by my own investigations, it seemed to amount to nothing more than the fact that gold had been found in the out-crops of certain *veins* of quartz, which had been so fully treated of, on pages 30 and 31, in Ch. V. of my Memoir; and as it had not been transmitted to the bureau, there was an opportunity to append a note to page 71, alluding to the additional confirmation it afforded of the views I had already given. Although it could not but be hoped that more correct accounts would soon follow, it was deemed proper to present, for the use of the reflecting portion of the country at least, the views in the note referred to, accompanied by a means of applying arithmetic in such manner as to demonstrate the utter absurdity of the story *as told*, and thus to aid in counteracting its mischievous tendency.

In place of the succeeding accounts having a tendency to correct the errors of the first, some of them were even more extravagant; and there continued to be published, for several months, such a constant succession of *auriferous quartz rock stories* that a general conviction of their truth pervaded, whilst it astounded the public mind.

It is believed, however, that the astonishment *was greatest of all among those who were actually residing in the localities of these imaginary auriferous rock formations*, into which the quartz or veins had been so wonderfully magnified. That I had ample reasons for believing the whole affair to be grossly exaggerated will be supposed by those who may have read the account of the gold region in Chap. I. and V.; and it was a source of regret that this work was in a position that prevented its immediate publication, because there were those who thought it possibly might have tended to induce some who were leaving their families and good occupations at home to pause before setting out in quest of treasures that few, if any, had the least probability of obtaining.

It is known to more than one member of the Senate, that after the Report was in possession of that honorable body, an endeavor was made to effect an arrangement, which, whilst it would have brought the work out in a more creditable style than that of our public documents, would have long since put it into the hands of its readers.

It is possible, however, that the publisher of my portion of the copies may be better satisfied that the delay has taken place, than if they had been issued six or nine months ago, when the received *aurimania* so generally prevailed. The facts, as observed and recorded by me, would perhaps have then appeared too tame and common place to suit the times.

So long a period has elapsed since the Report was written, that it seems proper to accompany its publication, at the present time, with these pages—the principal object of which, is to call attention to some of the views expressed in the Report, to point out confirmations thereof by subsequent events, and to notice a few of the erroneous impressions that have been formed by newspaper writers and others in reference to California.

In a practical point of view, it seemed so important to acquire a correct idea of the geological structure of the gold region whilst passing through it, that all the attention was paid to the subject that our rapidly performed reconnoissance would admit of. The facts, and my own conclusions in regard to them, are recorded in the Report itself—to which the reader is referred.

The official report of the Hon. T. Butler King, printed by order of the House of Representatives, differs so widely from my own, in reference to this and other matters, that it may be proper first to notice certain parts thereof.

On page 24, he says:

"The principal formation or substratum in these hills is talcose slate; the superstratum, sometimes penetrating to a great depth, is quartz. This, however, does not cover the entire face of the country, but extends in large bodies in various directions, and is found in masses aud small fragments on the surface, and seen along the ravines and in the mountains, overhanging the rivers and in hill sides, in its original beds. It crops out in the valleys and on the tops of the hills, and forms a striking feature of the entire country over which it extends."

This, as a description of the geological structure of the gold region, greatly surprised me, because Mr. King travelled with the same party that I did, through the portion of the region between the Yuba and Calaveras—and no indications whatever were observed by me of an approach to a structure of the kind he has given. The fragments of quartz seen on the surface in many places, (as often noticed in my Report,) were nothing more than the *outcrops of veins*, some of which, it was evident from the abundance of the fragments, must have considerable thickness. The real structure was sufficiently apparent, in my opinion, and was often a subject of conversation among the party; but I have no recollection that the idea of the quartz being disposed as a *super-stratum*, was suggested, if it had been even imagined by any one.

A few lines further on we are told that—

"The gold, whether in detached particles and pieces, or in VEINS, was created in combination with the quartz."

This extract was not made with a view of discussing the theoretic questions relating to the origin of gold, for the purpose of shewing whether the gold exists *in combination* with the veinstone, or is *disseminated* therein, but simply to aid in discovering the sense in which the writer uses the term "vein," to which we shall revert presently.

It is stated in the succeeding paragraph that—

"The rivers in forming their channels, or breaking their way through the hills have come in contact with the quartz containing the GOLD VEINS, and by constant attrition cut the gold into flakes and dust."

It may be readily supposed that this attrition would wear or rub off portions of gold in the state of powder, but it is not easy to imagine how a metal having the physical properties of gold could be *cut* into flakes by the means indicated.

On page 25, we are informed—

"It is beyond doubt true that several VEIN mines have been discovered in the quartz, from which numerous specimens have been taken, showing the minute connection between the gold and the rock, and indicating a value hitherto unknown in gold mining."

In this and in the preceding sentences it would seem to be indicated that the gold forms VEINS in the so-called superstratum of quartz, which we are told also contains the metal " in detached particles and pieces."

What is meant by " the minute connection between the gold and the rock," it is difficult to discern, unless we are to suppose that besides the gold veins and the detached particles and pieces of this metal, there is an additional quantity minutely disseminated in the quartz. This supposition is strengthened by the next observation, in reference to the specimens " indicating a value unknown in gold mining."

"Again, we are informed, "that the rivers present very striking, and it would seem, conclusive evidence, respecting the quantity of gold remaining undiscovered in the QUARTZ VEINS." "All of them (meaning the ravines through which the streams flow) appear to be equally rich," (which is inferred, by assuming as a fact,) "that a laboring man may collect as much in one river as in another."

In another paragraph it is stated:

"Hence, it appears that the GOLD VEINS are equally rich in all parts of this remarkable country."

We find the term QUARTZ VEINS in one of these last extracts, without any remark from which we can learn what idea is conveyed by it. In those before noticed, the writer spoke of GOLD VEINS in the quartz.

The term *vein*, as used in geology, as well as in mining, has a definite signification, which should always be carefully attended to—in order to avoid a risk of conveying wrong impressions.

When describing the gold region myself, it appeared so important for correctly comprehending the subject, that the general reader should become acquainted with the technical signification of the term, and with the character of veins, that I entered very fully into the subject, as the reader may see in Chap. V. I have there also explained why the fragments composing the outcrops of veins, when they consist of *vein-stone*, so little liable to disintegration as quartz, usually spread, more or less, upon the rocks that intersected them, so that where the veins are large, or exist in close proximity to each other, as is often the case, the surface becomes covered with these fragments over a considerable space. Appearances of this kind may possibly have suggested the erroneous idea of a *superstratum* of quartz.

The larger metalliferous veins that have hitherto been much worked, are found to extend to depths, (with decreasing richness, as stated in Chap. V.) greater than has ever pursued them, although he has penetrated one-third of a mile beneath the surface. They have certain general characters, whether they contain gold or other metals; and as the term *superstratum* gives an idea of them altogether erroneous, it was deemed proper to enlarge somewhat upon the subject.

The grounds upon which Mr. King seems to have founded his opinions in regard to the "unexampled" richness in gold of the quartzoze veins, or in his own opinion, the superstratified quartz rock, must therefore be deemed very inconclusive. He has very properly refrained from stating an average product of gold, as some others have done, although they should have known that no reliance should have been placed in the yield of mere specimens from an out-crop, and that very extensive workings must be effected before an approximation, even, can be attained in this regard.

The various accounts of the richness and abundance of the so called *gold bearing quartz* rock, induced many of our ingenious countrymen to quit their ordinary business and to apply their skill and talents to the production of improved machinery for grinding or crushing this material in order to separate the metal. The machines having been shipped *via* Cape Horn the parties themselves, singly and in companies followed after, by more expeditious routes.

Among those who consulted me there was one of much intelligence and mechanical skill who spoke of the completeness of his machinery for grinding, washing, &c. To him it was replied, that it would doubtless prove most effective for the intended purpose, but that was the least important part of the subject for him to consider, in reference to the propriety of going to California to put it into operation. That the large bodies of reported gold bearing quartz would prove after the outcrops were removed, to be nothing more than descending veins securely held between solid rocks, and that the cost of mining such was enormous, whilst the chances were almost wholly against their containing gold in proportion that would pay expenses. And further that the quartz rock stories, as well as many others frequently published, were wholly inconsistent with what I had observed in California. I pointed out other pursuits in that country which could not fail to be profitable, and recommended that he should either apply himself to them, or stay at home. But it seems, however, that what he had been told and read, backed by the specimens he had seen, indisposed him to regard any other than the brilliant side of the picture, so that he left for El Dorado. We may judge of the extent to which his expectations were realized by the following extract of a letter to his near relative, viz:

"No such golden rocks as described" "exist at all—all humbug." "This quartz rock has been tried and not an atom of gold found in it;" such a result might have been expected because there are many classes of veins in most metalliferous regions which are termed *barren or dry*, whilst others contain metals or their ores.

I have also been favored by a friend with the following extracts from a letter, written by another gentleman of Baltimore County, and who is well known and esteemed. He had been several months in this "golden quartz" district, when the accounts referred to were received there. After some sentences that, as before, are omitted, because of their harshness, he says: "I was astonished, as all others were, to read the accounts of the richness of these *veins* of quartz, and now let me tell you that for 100lbs. containing any gold at all, tons have to be thrown out not worth any thing."

"A company from Tennessee came up prepared with machinery to commence operations; they have broken up and returned home. Previous to their arrival, one came from Washington City for the same purpose, with all the necessary fixtures; they are here yet, but would give all they brought with them to be back again in Washington. There is gold in some of the rock, but the very heavy expense of getting it out renders an investment in that way any thing but desirable."

In a letter to another friend, he says: " There is gold in some of the rock, but to get out one dollar will cost two at least !"*

So far as I have information of a reliable character, there is yet no certain evidence of a single vein having been opened that will pay for the working, unless it shall turn out that the five feet vein of Col. Fremont on the Mariposa, is of the productive kind. The Col. himself is of the opinion that the quantity of gold obtained in the ravines intersecting it, would indicate such to be the fact. It is too soon, however, in my opinion, to hazard an estimate of the mean rate of production, because this can only be done after it shall have been extensively mined from one opening, or sufficiently so at several points in its course, distant from each other.

It is very possible that the mistakes committed by those who originated these accounts, were in part caused by assuming the proportions of gold in certain specimens to be an average of the whole *veins*, which some how or other they have imagined were extensive *rock formations*—a misapplication.which conveys an erroneous view of the facts. The specimens exhibited in this country also, aided considerably in giving credence to the stories themselves.

There were those that ought to have known better, who have observed, " there is no doubt of the truth of the accounts, because I have seen the specimens," as if they afforded any proof, whatever, of the mean proportions of metal contained in the veins.

In the note on page 71, before referred to, some calculations were made for the purpose of showing the quantity of gold that would have remained in a ravine from the destruction of the portion of a vein once extending across the space at present constituting the ravine. A vein only one yard in thickness was assumed, supposed to contain the lowest proportion of gold in the accounts referred to, say $2 50 to the pound, of , the veinstone. Let us now assume one cent's worth only of the precious metal to the pound, (a proportion altogether too small to be worked) and we shall have $43 20 per cubic yard. And as the whole destroyed portion equals 1,200,000 cubic yards, we should have for the amount left in the surface drift of the ravine from a single vein nearly $52,000,000 from which a small proportion should be deducted for that reduced to fine powder and carried off during floods by adhering to the passing drift.

It has been before observed, that the quartzoze veins are very numerous, and that some of them have probably considerable thickness. Many of these veins must contain gold in some quantity or other, as is

---

* A letter from a correspondent of the Baltimore Patriot, published since the above was written, says, under date of 1st October, 1850, among other things, " there have been agents of thirty companies at Mariposa during the season looking for locations, and, with few exceptions, have been disappointed. Some have left the country in disgust, and others have broken up; and having lost all their investments, are in a destitute condition."

evinced by the fact, that it has been found in all the principal, and in most of the secondary ravines of the district.

I have purposely assumed so little thickness for the veins, and that they extend in the most direct line across the ravines, (which few of them do) in order that those who think the results of the workings in the ravines, prove that the veins are rich without a parallel, may have all the advantage that the case admits of. If the proportions of gold and dimensions commensurate with some descriptions that have been published, had been assumed, the result would perhaps, have shown that many billions of gold should have been left among the drift matters of a single ravine, from one vein only.

Taking sixty millions for the whole amount of gold produced by the end of the present year, and which has been collected from ravines, whose aggregate lengths amount to several thousand miles; and within which the workings have been prosecuted with great activity, during three seasons, with a force rapidly augmenting, until it has reached *more* than one hundred thousand men; and we must see that we can gather no hope that most of the veins will prove any better than those of other gold regions. That there may exist however, veins to be hereafter profitably worked, is very possible.

The anxiety to discover the rich workings, that most persons expected, made the gold-seekers of last year a roving people, incessantly moving about and digging holes in the drift, by which the richer deposits were discovered and first worked. The best of the leavings of former years, were again hunted out and first worked this year.

This process must continue during each season, and the richer places be successively exhausted, whilst the progressive reduction of the cost of labor, and the necessaries of life, will permit the working of poorer places successively.

In all extensive gold regions that have been examined and accurately described hitherto, it has been found that there is a remarkable similarity in mineral composition and structure, and that whenever these are met with, one may with great probability, if not certainty, indicate the existence of gold.*

The large amount of gold produced in California within the past three seasons, furnishes no ground whatever for the belief that supplies of similar amounts must be continuous. On the contrary, by such rapid working, these thin deposits of drift must be the sooner exhausted. This conclusion cannot but force itself upon the mind of every one who will carefully apply the actual facts bearing upon the question.

* In the year 1837, in the Trans. of the Md. Acad. of Science, it was suggested by me that the talcose slates of Maryland might be expected to contain gold, and about two years since a vein was opened in Montgomery County. A few years later, Mr. Dana also indicated that the slates of the northern part of the Sierra Nevada, would prove auriferous, as they have done.

When the reader shall have read the descriptions of the Topography and Geology of the gold region in this work, he may remember that it is deeply—very deeply—indented by numerous principal and secondary ravines, which, in most cases, are extremely narrow at the bottom.

We have seen how small a proportion of gold in veins (of which such extensive portions have been worn through and broken up,) is requisite to leave a large amount of the metal among the drift principally in the bottoms of these deep ravines. In the *small* ravines of our Atlantic gold region, we could only expect a quantity of metal proportioned to their size, if the veins were, in all regards, alike in both districts.*

The deep ravines in Peru, New Grenada and Chili furnished large supplies of gold at first, but those most attractive were soon exhausted, and the supplies from them, of course, were rapidly reduced.

The gold regions under the dominion of Russia, cover a much larger area than we have west of the Sierra Nevada; and the supply has steadily increased in amount until it has reached $20,000,000 per annum. The workings are systematically prosecuted, under police regulations, and the operatives are mainly convicts and the serfs of the nobles. The veins, in almost every instance, have proved unproductive, and are rarely worked at all. Extensive as these regions are, they would have long since been exhausted, but for the inhospitable climate and the unpopularity of the government of the country.

It is evident that the little width and depth of the strips of alluvial drift, in the deep ravines of the Sierra Nevada, caused the gold to be deposited in narrow lines in their bottoms; so as to enable operators to exhaust them with much greater rapidity than if it had been diffused among, or covered by that of wider valleys.

The extraordinary numbers who have flocked to California since the existence of the gold was made known, have, it is believed, occupied almost the entire length of every gold producing ravine, of any consequence, from Oregon to the Maripoosa at least,—and have, in a considerable degree, exhausted the gold from those locations, that could be worked at the value of labor up to the present time.

If, as is generally supposed, the white population of the country amounts to 200,000, there must be little short of 150,000 who are compelled to employ themselves at gold digging for their daily bread; because, it is not probable that the resources of the country are yet sufficiently developed to afford employment and support to more than 50,000, including mechanics, traders and transporters. The urgent calls of hunger will compel these unfortunate people to pursue their calling with the utmost perseverance, although the reward for a given

* We should not, therefore, be too hasty in believing the veins of California to be richer in the precious metal than those of our older States, in which there are veins that have been and continue to be mined, it is believed, with advantage to those concerned; although the gold in the ravines was very soon exhausted.

amount of toil progressively lessens. And this state of things must continue for some time longer, or until other branches of industry can be sufficiently expanded to absorb most of the misapplied labor.

From what has been already said, in connection with the actual character of the auriferous surface deposits, the supply of gold could not be other than large during the present, and perhaps another year; but must inevitably lessen, very materially, afterwards.

But for the delusive accounts of the gold bearing quartz rock, before noticed, the immigration into California, within the last twelve months, would not have been large, and all the better for the country, because a moderate number of new comers might have found employment in the various branches of industry, which the valuable resources of Oregon and California are giving rise to.

Those who have read my Memoir will have seen that such a vast influx was not at all anticipated by me, because, there was no state of *facts* presented to my view that could justify it. It would have been difficult to have imagined that such an extravagant series of exciting stories would have been made out of the real facts, or that they could have obtained almost universal belief among our people.

In modern times, at least, there is no instance of a gold region equal in size to that of California having been worked with such extraordinary rapidity, and which, therefore, must so soon show symptoms of exhaustion, so far as regards the gold in the drift, which being near the surface is accessible to individual operators.

Unmistakeable evidence of these symptoms appeared in many of the secondary ravines especially, which we traversed during the summer of 1849; and even before the close of the working season, the larger ravines presented similar indications throughout much of their extent, as I was credibly informed before, as well as since my return.

Another strong indication that most of the diggings in the central portions of the gold region, (between the Yuba and the Calaveras) are becoming less productive, exists in the fact that the accounts received during the present season seldom refer to operations therein, although it was from this portion of the region and that about the Stanislaus, that most of the gold was obtained during the two preceding seasons.

The more northern and southern portions of the gold region, seem to be most in favor at present, but are worked by such numbers as must reduce them in a shorter time, to the state of the middle portion.

It is sufficiently evident according to the information I have been able to collect to the present time, that the actual state of the diggings is such, that few can procure more than the absolute necessaries of life, so that being unable to engage in pursuits requiring capital, most of them must continue at digging until the increasing demands for agricultural and other labor, shall absorb all that may be unprofitably employed in the gold region.

This very state of things, by keeping the diggers poor, causes the proportion of the latter to be too great for the number of those who supply their various wants, and whose profits are therefore inordinately great, because of the comparatively little competition. If a reasonable proportion of the former, could accumulate sufficient capital to enter the better class, the matter would soon regulate itself. Unfortunately, however, a very small trading business in the mountains, requires an amount of capital that would surprise an Atlantic dealer.

The amount of labor employed in digging for gold, compared with that of other branches of industry, even last year, was altogether too great for the true interest of either California, or the older States; and it is unfortunate that events occurred to increase the disproportion. If the talents, means and labor already misapplied in preparations for mining and grinding the quartz, had been devoted to agricultural and other pursuits adapted to the country, it would have been better for all parties. As these pursuits are commenced or extended, relief will be afforded to surviving gold seekers and miners, by absorbing a portion of the misapplied labor, as well as by lessening the cost of supplies, so that eventually, ravines previously worked, in the unartistic manner pursued last year, as well as places too poor to be operated upon at the present time, may perhaps, be worked over again with labor at a moderate cost, aided by skill and capital, systematically applied. Under these circumstances, we may, after another year or two, have moderate and steady supplies of gold for a long period, that will cost no more than it is worth.

From what has already been said, it would seem about as prudent to infer, that a spendthrift, because he *has* spent large sums of money, must *still* be wealthy, as that California must continuously produce great supplies of gold, from easily accessible surface deposites, worked with such unexampled rapidity.

In these observations relating to supplies hereafter, those from the drift only are taken into view, because the prospect of a material addition to be obtained by *mining* the veins, is altogether too remote and uncertain to be relied on.

If we are willing to profit by ample experience in both hemispheres, in this regard, as well as by our present knowledge, limited as it is, of California, we might conclude, that if veins be discovered that can be advantageously mined, little could be got from them for some years, because of the large amount of capital required for labor, machinery, &c., as well as the great risk of loss in such mining operations.

The reader may perhaps recollect that numerous statements have been made in the accounts from California during the past two years, of the discoveries of many large pieces of gold, whose weights were asserted to be from 90 lbs. downwards. As such specimens are *rather* costly for most persons to retain, it is probable that few such remained uncoined.

In September last, the papers announced the receipt of a lump, at

Newark, N. J., weighing 23 lbs; and a few days afterwards, it was stated also that there was one of 53 lbs. at Boston. Whether these were solid, *native* metal, or had been produced artificially, by melting up quantities of the gold, as ordinarily met with, or whether they were pieces of quartz, containing fragments or scales of gold, we are not informed. The first large piece that reached the U. S. was brought in by Lt. Beale, U. S. N., about eighteen months since,—and it was said that some, of somewhat greater weight, had been previously sent to England by the British Consul at San Francisco. Of the numerous *reported* lumps, but four had reached the U. S. Mint on the 30th of October last, viz:

By Lieut. Beale, ...	6 lbs. 9 oz.	Troy wt.
" Mr. Mickle, . . .	9 " 7½ "	"
" Mr. Perkins, ...	14 " 6 "	"
" Mr. Post, ...	5 " 7 "	"

It would be a matter of interest perhaps, (*if any others exist*,) to know where they are.

For the sake of comparison, the weights of large lumps, heretofore found in other parts of the world, are given as follows:

At La Paz, in Peru, in 1730, ... 59 lbs. Troy wt.

Raynall says this was not uniform in composition, but varied in the proportion of gold, from 75 to 95.8 pct.

In Sonora, Mexico, (Humboldt) .	48 lbs. 6 oz.	Troy wt.
In Siberia, according to Humboldt,	27 "	"
(Deposited in the Imp'l Museum.)		
And others, respectively weighing		
26.25, 20, 16.64, 14.25, 11 lbs.		
In Cabarras Co., N. Carolina, in		
1828, (Reid's Mines,) . .	28 "	"
And another, . . .	13 "	"

There appears to be more uniformity of composition in the gold of California, than in that of other regions which have been sufficiently examined for comparison—as will appear from the following table of results from the most important localities:

	Maximum.	Minimum.
California, ...	95.70	81.20
Siberia, ...	98.96	28.
South America, (New Grenada, &c.) .	88.58	64.93

Many of the pretended estimates of amounts of gold exported, or taken from California hitherto, approximate the truth, perhaps, as near as some of the stories before adverted to. One of these, which went the round of the papers in September last, made out the sum to be $150,000,000! This, I recently reviewed at some length in the Baltimore American, and made such corrections as reduced it to $43,000,000; although in several of the items, I had strong doubts that sums too large were admitted. This *great* estimate sent $45,000,000! to England by different channels! and yet by the last steamer, the papers give us "the striking piece of intelligence of the arrival, at Southampton, of one million of dollars worth of California gold dust!" Now, there would be nothing so "striking" in this fact, if it were not a very rare occurrence. England has furnished directly a small portion of the imports into California—and is therefore entitled to her share of the gold, as remittances; but as by far the larger amount of supplies went from our Atlantic States, the proportionate returns in gold necessarily must come here, and find their way into the U. S. Mints. The assertions so often reported, of large sums having been carried off by Mexicans and others, have about as little foundation as the gold bearing quartz stories, and no more.

The amounts received at the Mint are as follow:

To December 31st, 1849, (at Philadelphia,) . .	$5,525,616
" " " (at New Orleans,) .	666,079
	$6,191,695
From January 1st to September 30th, 1850:	
At Philadelphia, . . . $19,271,463	
At New Orleans, ... 2,495,248	
	21,766,711
Aggregate to September 30th,*	$27,958,406

To which is to be added what may have been used in the arts, or retained as specimens.

The announcement of the arrival of a steamer from Chagres, has been too frequently accompanied by exaggerated notices of the value of the gold brought by her. Even, when the first notice of the amount on freight is correctly stated, we were often told that there is "as much more in the hands of the passengers." Upon my return, we had $625,000 on freight on the passage to Panama, and about the same sum from Chagres to New York; and, as far as could be learned from the passengers, it was not believed the amounts in their hands much, if any, exceeded $100,000, and yet the *first* announcement in the papers was that there were "one million of dollars worth on freight, and as much more in the hands of the passengers!" The correct freight list was subsequently published. Any one who has carefully compared the amounts brought on freight to Panama, with those received at New York and New Orleans, will have perceived how closely they approxi-

---

* A letter from the Director of the Mint, received since the above account was furnished, says: "The total amount of this bullion received at the mints, from the discovery of the mines to October 31st, 1850, was $31,838,079. The deposites of the current month (November,) will probably advance the total to thirty-six and one-half millions.

mate. The small proportion transferred to the British steamer at Chagres, is perhaps about equal to that received on board our steamers at Mazatlan and San Blas, from the mines in the north-western part of Mexico. The freight lists of the steamers, by approximating so closely the receipts at the Mint, prove that nearly all the gold is sent there for coinage. A small portion is brought by passengers, but this is a violation of the rules of steamers, and large amounts cannot be thus transported, without detection and paying the freight.

There are houses in San Francisco having open policies of insurance, and who deliver the gold in New York, free of all risks; and prudent persons generally make use of them for large sums.

Having thus disposed of the subject of gold in connection with California, some other matters of interest will be briefly noticed.

In treating of the agricultural products of the country in Chap. XI. attention was directed to the subject of horned cattle, whose hides and tallow formed the principal staple for export, before the golden era. It might have been supposed that there was danger of such a sudden influx of population producing a scarcity of beef within a short time; but information derived from those well acquainted with the whole country west of the Sierra, indicated that there was no danger to be apprehended in this regard; and that the natural increase (so rapid in that climate,) from the *immense* herds in the country, would continue to furnish ample supplies of beef, for all the population that would have been expected during succeeding years.

Mr. King has given a different opinion, founded upon an estimated immigration of 100,000 persons per annum. It is possible, if the immigration be continued at this rate, that the rate of consumption of beef may, in a few years, exceed that of the increase. But as it is not easy to imagine exciting causes for each year, equal to that of the present, we cannot suppose a similar influx hereafter,—and indeed it is not unlikely, that within the next twelve months, nearly as many may return home as will go out; although, afterwards, the population of the country will steadily increase.

On p. 14 of Mr. King's Report, he gives estimates to prove that the consumption of beef in California will, by the year 1854, require all the stock then on hand, and the increase to the period named.

On the same page, he says—

"Cows are driven in considerable numbers from Missouri and the time cannot be far distant, when cattle from the Western States will be driven by tens of thousands to supply this new market."

"If California increases in population as fast as the most moderate estimate would lead us to believe, it will not be five years before she will require more than one hundred thousand head of beef cattle, per annum, from some quarter, to supply the wants of her people."

On page 15, he says—

"I am acquainted with a drover who left California in December last, with the intention of bringing in ten thousand sheep from New Mexico. This shows that the flocks and herds east of the Rocky Mountains, are looked to already, as the source from which the markets on the Pacific are to be supplied."

There is little doubt that a drover will do well with his *flocks* of sheep if he is successful in getting them to California, for the reason as was stated in Chap. XI. page 58, that sheep are well adapted to the country; and there is a great deficiency in the supply, because the attention of the rancheros has been turned hitherto almost wholly to horned cattle and horses, whilst sheep are most abundant, and can be purchased at a very low price in New Mexico.

So many of the inhabitants have abandoned the gold diggings and turned their attention to agriculture, that there must, for some years to come, be a great demand for sheep, to assist in stocking their farms, as well as for present use, and which no doubt the discerning drover determined to profit by. But it by no means follows, that the drover's projected speculation in *flocks* of sheep, shows that *herds* of cattle will also be required from abroad.

Before we take leave of subjects connected with agriculture, a few words may be said in reference to the *Burr Clover*, an important element for pasture in California. I expressed some doubt in Chap. XI. page 61, as to its being clover, as I had collected imperfect specimens and those even failed to come to hand, but having been favored by a fellow-passenger with a few seeds which grew finely, I had an opportunity to observe that it is a true clover, and is at least worthy of trial on this side the Rocky Mountains.

In treating of lumber, Mr. King observes on p. 21 "that the price had recently fallen at San Francisco to $75" and that "at that price it cannot be made where labor is from $10 to $15 per day." When I left San Francisco in October, 1849, there was no difficulty in procuring American and European labor at from $6 to $8, and that of the Spanish and Indian races at a much lower rate. Soon afterwards the value of labor was reduced very much lower, and although some improvement in price took place as summer approached, the great influx of people into that country and Oregon during the present year must necessarily ere this have reduced the cost of labor very considerably.

Without desiring to trouble the reader with many figures, reasons and facts, he may be referred to p. 41 and 44, where it was remarked that there is an immense amount of the finest timber, easily accessible both in California and Oregon. Even now it can, in my opinion, be delivered at the tide water ports at much less cost than $75. Henceforth the price must, in general, continue to decline for some time to come; so that little or none will be shipped from the Atlantic hereafter, without loss, except such as may be wrought by means of machinery into various forms applicable to house building and other purposes. Even shipments of these will not long continue, because our countrymen will soon erect the requisite machinery on the borders of the Pacific, and put a final stop to all importations of lumber from the Atlantic.*

Mr. King estimates the cost of lumber in our large markets at $16, and the freight to San Francisco at $24; he adds also "$20 to meet any unforseen increase of price, or in the freight" and supposed, that this state of things would "cause the manufacture of lumber in California to be abandoned." The cost of delivering lumber from the Atlantic, (here assumed) appeared altogether too low, but instead of presenting the result of my own investigations, I shall give those from a well known journal, published at Portland, Me., where the lumber trade, in all its relations, is perhaps as well understood as at any place on the globe. The extract is from the Baltimore Sun, in April last.

" *Lumber in California.* — The Portland Advertiser, alluding to Mr. King's late report on California; says: Our largest foreign market for boards has been the West Indies. Voyages from Maine to those islands are performed in 12 to 25 days, and our merchants, we are informed, consider $5 to $6 low freight for boards from here to the West Indies. Now, allowing an average of 20 days for the voyage to the West Indies, and 5 to 6 months for the voyage to San Francisco, the above rates of freight would give us $45 to $50 per M. as a freight, only equivalent to what is considered a low freight to Cuba, &c. When we take into consideration the fact, that there are no return cargoes at California of consequence, and the risk of desertion by the crew—even $50 is not so good as the usual freights to the West Indies. Even at $100 per M. there will be but few shipments of boards to California from the United States, for the expenses there generally exceed those of any other place, to which lumber is now exported. The late news has checked shipments, and unless more favorable accounts are received, but few cargoes will go forward for some months to come."

In Chap. VIII. the subject of lumber was considered, and from the information received before I left the country, in reference to the rapidly increasing number of saw-mills, there appears little doubt, that, within a very short period, our Pacific possessions will not only supply their own demand for lumber, but when the existing excitements shall have subsided, a large export-trade in this article will originate in California and Oregon, which, at no very distant period, will greatly add to their commercial importance.

* Since the above was written, we have accounts from San Francisco, with price currents and reviews of the markets to the 1st Oct. The following relating to the prospects for a lumber trade from the Atlantic States, is extracted, and would seem to fully confirm the views given above, as well as those on page 41 of the Memoir.

Lumber—Notwithstanding the large quantities consumed—causing a fair demand—we have no general improvement to notice—some kinds doing a little better—timber used for wharfing and laying the streets being most saleable. *It is supposed that lumber from the States will not, except a very few kinds, ever pay in this market,* as large preparations have been made for sawing and preparing lumber in Oregon and other places on the coast and islands. Even house-frames have been used in laying the streets. It is allowed that we have now a year's stock if another foot should not arrive.

The demand for flour it is estimated by Mr. King (p 22) will amount to half a million barrels, when California shall contain 200,000 people, and which is supposed to be the number in the State at this time. He is of the opinion that "no country can supply it so good and so cheap as the old States of the Union," and adds "there is no pretension to accuracy in these items, they may be estimated too high; but it is quite as probable they are too low." The consumption in our Atlantic cities is believed not much, if any, to exceed one barrel a year for each individual, to which may be added the small amount of flour from other grain used for human food. We might however, admit fifty per cent. more for California and yet the 200,000 persons would require only 300,000 barrels, besides barley and other grain for domestic animals, of which however, large amounts will not probably be required before the country will be in a condition to produce them in abundance.

It is well known that Chili is a fine grain growing country; the produce thereof has been steadily increasing during the last twenty-five years; and now more rapidly in consequence of the excellent market in California. Although this sudden demand could not be fully supplied, there is every reason to believe that in a short time it will meet all wants in this regard that cannot be supplied in California and Oregon. Besides the day is rapidly approaching when the State and Territory together, will furnish such ample supplies of breadstuffs as to put an entire stop to such imports from abroad.

The observations of Mr. King in reference to the *consumption* of coal seem to be justified by present indications, at least as far as relates to the use of it by steamers, but he thinks " if it be delivered for $20 per ton " at San Francisco, (which it cannot be at the present time in sufficient amount, if information given me by shippers be correct) that " it would compete successfully with the coal from Vancouver's Island and New Holland." I am not sufficiently acquainted with the position, quality, &c., of that from the latter country to speak of it, and regret so little is known of the former. It is however known that the coal of Vancouver's Island reaches within a few miles of good harbors for vessels; and *specimens* which I have examined and seen burned indicated a quality for steaming purposes nearly equal to that of good English coal shipped to this country and the Pacific; although inferior to the best of the coals of Western Maryland which have hitherto reached tide water. A trial of it on board a British vessel also proved that it was well adapted to steaming purposes. The thickness of the veins, the character of the roof and other circumstances that must be known, in order to estimate the cost of elaborating it, I am not informed of, but it would be most unfavorably placed in these regards, for bituminous coal, if a man cannot mine two tons of lump coal per day. There are many mines in this country and in Europe wherein three or four tons per man are daily mined. But we will suppose Mexican underground miners to be employed, and that but one ton a day be produced by each, at the high cost of five dollars, although an investigation of this *particular subject* satisfied me that any number of such miners could have been procured from Mexico at a much lower price. If the cost of conveying the fuel to the vessels (nine miles as I am informed) be as much at the present time as five dollars more, including profit, and the cost of freight to San Francisco be even five dollars, for the ten days voyage down, and thirty back, the coal could be sold at San Francisco for fifteen dollars, exclusive of commission. And yet all these items of cost are, in my opinion, very much over rated.

The present owners of this coal are the proprietors of the Hudson's Bay Company—too unwieldy a concern for prompt action in such a case; but, it is not to be supposed they will permit a treasure so valuable to remain long idle. An English paper not long since, (I think the London Spectator,) stated, that means were being taken to work the Vancouver coal without further delay. When they are made fully effective, it is more than probable that this coal will be delivered in San Francisco, at less than half the price it can be shipped from the Atlantic ports.*

There are also other coal formations on the N. W. coast of America, both in the British and Russian possessions, that will soon compete with that on Vancouver's Island, and lessen its price.

---

* The reports, other than my own, reached me since this Introduction was written. In that of Gen. Smith, page 86 of this work, we are informed that the Hudson's Bay Company will deliver coal at $14 per ton. This is, of course, *certain* information that the cost of coal will, for the present, be $1 less than I have estimated.

Among other localities in Oregon in which coal has been stated to exist, it was announced some two years since, that it had been found on the Celeetz River. In the Report of Lieut. Talbot, page 110, he informs us that he examined this locality, and "found a seam of 4in. thickness;" and is of opinion "that larger seams of coal must exist in the vicinity."

He adds, that "specimens of this coal have been submitted to practical miners and others, who pronounced it to be anthracite, of good quality."

A specimen of it appears to have been analyzed by Professor Frazer, of Philadelphia, and he considers it to be lignite. See page 117.

This is one of the cases that serve to illustrate the observation on page 3, of this Introduction, as to the frequency of mistakes in reference to minerals. The *practical* miners and others "pronounced it anthracite, of good quality;" whilst the analysis and examinations of Professor Frazer *prove* it to be lignite, which differs *so widely* from anthracite in all its characters, both physical and chemical, that little knowledge of fossil fuel is requisite to distinguish them.

Mr. Frazer has (no doubt inadvertently,) made an observation, in reference to this coal, that might occasion groundless expectations; he says " that it occurs so near the deposites of Vancouver's Island, as to give rise to the hope that it may be a part of the same deposite."

The geological position of the Vancouver coal has not been described, so far as I know of—but the specimen obtained by the proprietors of the Pacific Mail Steamship Company, and the analysis made in England, both show that it is from a *true bituminous coal* formation—and, of course, it belongs to a geological period, vastly older than that of the tertiary lignites, which, as was before stated, are distributed on the Western coasts of America in numerous places, from Oregon to the Straits of Magellan. Even if the Oregon lignite were a *true coal*, its distance from the known coal formation of Vancouver's Island (in lat. 50°) is at least 300 miles; so that no hope can be derived from their supposed proximity, and besides it would seen from notices of the intervening country, in the Geology of the Exploring Expedition, that the two localities are separated by igneous rocks.

It was only designed to refer to those parts of the report of Mr. King, relating to the productive industry of the country, but it seems proper to notice his views, in reference to the prevailing winds of California, which are the principal cause of its peculiar climate, so different from that of our Atlantic States. On pages 8 and 10 he states, that these winds are from the *Northeast*, whilst Fremont, Wilkes and others who have written about them, tell us, they are from the *Northwest*, which was certainly the case during my sojourn in the country. I have so stated in Chap X., wherein the interesting subject of climate is considered somewhat at large. The best informed persons, that I met with, who had resided some years in the country, testified to the same effect, adding also, that *southwesterly winds* often prevailed in winter, and that there were occasionally during that season severe gales from the southeast. The theory, by which Mr. King accounts for the northeast winds, he speaks of, is wholly inapplicable to those from the Northwest, and unless I am greatly mistaken, is the one usually applied to the trade-winds, prevailing in the tropical regions to about 28" north latitude (and not even so far north, on the eastern shores of the Pacific.) They, of course, reach no part of Upper California.

It would have been far more pleasant to have been able to coincide with Mr. King, than to differ with him in so many important particulars.

His report has been widely circulated, and whilst it contains much useful matter, it would seem due to the public, that its errors, if errors they be, should be pointed out.

It is a matter of great importance, that the community should have the fullest means of getting at the actual condition of affairs in California.

The course of events will very soon show, whether the great expectations that have been entertained by the public, relating to the supplies of gold from California hereafter, are likely to be realized, or whether the results will accord more nearly with what some others, as well as myself, have endeavored to indicate. Under the impression, that the digging for gold had absorbed an inordinate proportion of the labor of immigrants, my report was, in part intended, to bring into notice some of those important resources, which constitute the real and permanent wealth of the valuable portion of our country, washed by the Pacific. If the means, that have been in part adopted, have hitherto been inoperative from the delay in publication, the fault lies not with myself.

On the present occasion, means have been taken to prevent any additional delay by putting my portion of the copies into the hands of an enterprising publisher, by whom it will be issued, in connection with the additional matter, now furnished him.

A brief stay in California, less than four months, served to create in my mind a lively interest in reference to that country, without however presenting opportunities for learning as much of it as would have been desirable. I could not but deeply regret that the state of the country and other causes, prevented further investigations and extending them through other portions of the country as well as to Oregon; because this territory and the auriferous state, although separated politically, are so closely bound together by industrial interests, that they can only be considered as one and inseparable. Their resources, varied as they are, may be extensively developed in the progress of time, but as there is so large a quantity of *public land* in those regions, it may be supposed that means should be adopted by the general government, to cause such investigations to be made as may tend to make known the existence of every thing calculated to promote the commerce and all other branches of industry. And although I shall not be one of the explorers, I may be permitted to assure those who may thus occupy themselves, that they will find not only a most interesting field for research, but one in which they may do much to extend the empire of the great Anglo-Saxon family, in a manner more useful and more humane than by the sword.

It may be safely asserted that no new country ever received as suddenly, such a large population so well fitted in every regard to found a great nation, as have poured into it from the States; and though we may lament the distress, ruin and death that has been brought upon so many of our countrymen by inconsiderate publications, yet we may be permitted to hope that under a wise providence, great and beneficent results will be produced by this excessive immigration to the benighted shores of the Pacific.

The mania for gold digging with them is more and more being succeeded by the desire to engage in other pursuits, which are already producing good results in rendering the inhabitants less dependant upon other regions for the necessaries of life, whilst occupations are afforded not only more healthful, but more likely to give certain returns to the parties themselves.

As I have elsewhere observed, the Spanish races are noted for the pertinacity with which they continue to work in the diggings and mines of the precious metals, although they may be far less remunerative than other branches of industry. In Mexico, for instance, well informed persons inform us that if the whole mining labor were applied to other branches of industry, the value of its products would be more than double that of the gold and silver obtained. These races seem continually to be led on by hopes of their workings being more productive, the same sort of delusion that encourages the gambler or adventurer in lottery tickets.

The historian Robertson, asserts, that when they have entered this dangerous career, their infatuation permanently chains them to it. Can it be supposed that our energetic and calculating race will fall into similar habits? We cannot think they will; but, on the contrary, we feel assured that, true to their natures, our people will generally continue to apply thmselves to those branches of industry most likely to give a *certain* reward for their labors.

California and Oregon possess a vast expanse of fertile soil; which, with their varieties of climate, will enable them to raise all the necessaries and luxuries of life that either Europe or the United States can produce. And these in abundance for a dense population, and to spare. But for the little hope that their geological constitution (as far as known) gives us to expect the existence of *true* coal, we should say that they have facilities also for manufacturing on a great scale; and even if this deficiency shall continue, it may in some measure be made up by the large coal fields of Vancouver's Island, and the adjacent continent, sufficiently convenient to the navigable portions of Oregon, and California.

The unproductiveness of gold digging is fast curing the mania *there* at least, and as it declines, other branches of industry will attract more attention. Finally, gold digging or mining will be prosecuted only where they will pay as well, or better, than other pursuits, and each branch of industry will assume its proper place in the general system. With so large a capital of mental and physical force to begin with, we cannot but anticipate that a wide spread commerce will result, and that a great Anglo-Saxon nation will rise on the Northern shores of the Pacific, to produce a mighty influence upon all its borders.

BALTIMORE, 10*th Nov.* 1850.

––––––––––

RESOLUTION OF THE U. S. SENATE.

Resolved, That the Secretary of War furnish the Senate as soon as practicable with any recent Report, or other information, in reference to the Geology and Topography of California, now in the possession of the War Department.

REPORT

OF

THE SECRETARY OF WAR,

COMMUNICATING INFORMATION IN RELATION TO THE

Geology and Topography of California.

MAY 6, 1850.

Ordered to be printed, and that 5,000 copies, in addition to the usual number, be printed for the use of the Senate, 1,000 of which for the use of P. T. Tyson.

WAR DEPARTMENT,
Washington, March 28, 1850.

SIR: In compliance with a resolution of the Senate of the 25th ultimo, requesting to be furnished, "as soon as practicable, with any recent report or other information in reference to the geology and topography of California now in possession of the War Department," I have the honor to transmit herewith copies of reports from Brevet Major General Smith and Brevet Brigadier General Riley, with accompanying papers; also a communication from the colonel of the corps of topographical engineers, with a copy of a memoir upon the geology of California, addressed to him by Philip T. Tyson, Esq. The report of General Smith, and the accompanying papers, contain also interesting information relative to Oregon.

Very respectfully, your obedient servant,
GEO. W. CRAWFORD,
Secretary of War.

Hon. M. FILLMORE,
President of the Senate.

ERRATA.

The reports which follow, not having the advantage of the author's revision, errors too important to be overlooked have occurred; and which should be corrected before perusal.

On page 8, line 12 from top, for "ruins" read *veins.*

" 11, " 5 " " for "from" read *than that of.*
" " " 24 " bottom, after "serpentine" add *and its allied rocks.*
" 12, " 5 " " for "drift" read *dip.*
" " " 3 " " for "Dalen's" read *Daler's.*
" 14, " 16 " top, for "basing" read *basin.*
" 16, " 23 " " for "Treusum" read *Suisun.*
" 17, " 12 " " for "brimstone" read *limestone.*
" " " 19 " bottom, for "range" read *ridge.*
" 22, " 18 " top, after "to" insert *two hundred and fifty.*
" 26, " 5 " " for "thin lime" read *their lime.*
" 27, " 23 " " for "on" read *or.*
" 27, " 12 " " for "these" read *their.*
" " " 7 " bottom, for "western Asia" read *eastern Asia.*
" 28, " 8 " " for "vein" read *veins.*
" 29, " 13 " top, for "river" read *vein.*
" 32, " 1 " first line, for "where" read *whose.*
" 34, " 18 " top, for "on" read *along.*
" 47, " 25 " " for "coast" read *east.*
" 51, " 15 " " for "west" read *east.*
" 54, " 6 " bottom, for "fig. 9" read *pl.* 8.
" 55, " 30 " top, for "the basis" read *their bases.*
" 63, " 24 " " for "calium" read *calcium.*
" 65, " 6 " " for "fig. 7" read *fig.* 2, *pl.* vi.
" 66, " 5 " bottom, for "fig. 8" read *pl.* vii.
" 69, " 25 " top, for "western" read *eastern.*
" 70, " 6 " " for "is" read *are.*
" 71, " 6 " bottom, for "north" read *south.*
" " " 6 " "1000" read 5000.
" 73, the letters of reference to plate VI. are wanting on the plate, but the reader will easily find the places for them.
" 116, line 3 from bottom, for north read *south.*

The word "Septenite" occurs in several places instead of *Leptinite.*

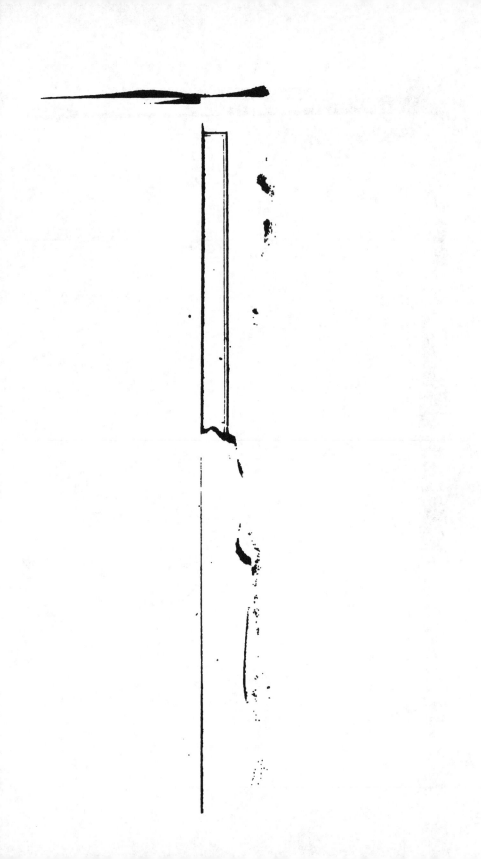

Report of P. T. Tyson, esq., upon the Geology of California.

. BUREAU OF TOPOGRAPHICAL ENGINEERS,
Washington, March 26, 1850.

SIR: In compliance with a resolution of the Senate of the 25th of February, I have the honor to submit a copy of a report upon the geology of California, from Mr. P. T. Tyson, of Baltimore.

This gentleman, who is well and extensively known in our country for his geological attainments, visited California in 1849, for the sole purpose of investigating its geological peculiarities. When he returned, he furnished this office, at my request, with a report of his labors, the copy of which is now submitted.

Respectfully, sir, your obedient servant,
J. J. ABERT,
Colonel Corps Topographical Engineers.

Hon. G. W. CRAWFORD,
Secretary of War.

———

. BALTIMORE, *January* 30, 1850.

DEAR SIR: I have the honor to transmit herewith a memoir upon the geology and productive resources of parts of California visited by me in the last year.

I had commenced preparing it before I learned you had expressed a desire to have it reported to the Topographical Bureau; but, during its progress, the subjects have been treated in such manner as I deemed would be most useful in connexion with the valuable information that emanates from the scientific branch of the public service over which you preside.

Besides the obvious propriety of giving the information to the public through the Topographical Bureau, I do it the more willingly because of the desire you expressed to me, before I left for California, that you might be enabled to add geological surveys to those made by the topographical corps.

In recording the results of my hasty reconnoissance, the primary object has been to aid in giving definite and accurate knowledge (to the extent of the means for collecting it) in reference to a portion of our country which seems to have furnished ample occasion for the exercise of what the phrenologists would call the organs of ideality and wonder.

The matters are mainly considered with reference to their industrial applications. Their scientific relations are for the most part embraced to an extent sufficient to illustrate the subjects treated of. It is not improbable but that these may form a separate paper, after the specimens which I collected shall have arrived and been examined.

There are numerous subjects of great interest in California to be investigated in their scientific as well as their industrial relations, but the cost and inconveniences attending them are so great as to render such investigations by private means very unsatisfactory.

I must acknowledge my indebtedness to several officers of the army and navy for facilities offered and furnished whilst pursuing my examinations in California.

Upon my arrival at San Francisco in June of last year General Smith and suite, accompanied by Mr. King, were about to make a tour through portions of the country, and the General kindly invited me to join the party. But for this I should have accepted an invitation from the lamented Captain Warner, of the topographical corps, upon the expedition wherein his devotion to the duties of his profession cost him his life.

I was much gratified to learn that the results of his valuable labors in California were safe. They will add largely to the amount of accurate information relative to the "Far West."

Very respectfully, yours,

PHILIP T. TYSON.

J. J. Abert, *Colonel of Topographical Engineers.*

—

Preface.

The object of this memoir is to present a brief sketch of the geological structure and industrial resources of parts of California, including those sections of the country which have recently attracted so much attention.

It is likely these sheets would not have met the public eye but for the expressed desire of friends whose opinions are entitled to consideration, who thought that information of a more precise and definite character than is yet before the public might be useful to the community.

Personal observations were made in the summer and early part of autumn of the last year, and, it may be said, for the most part during journeys too rapidly performed to take other than hasty glances of the country which was travelled over. It is true that better opportunity offered at some points, but these in many cases were not in localities where minute investigations would have been most desirable.

Persons not familiar with the circumstances affecting the characters and positions of rocks and mineral matters in general are so prone to be deceived in their perceptions in regard to them, that little aid to be relied on was to be obtained by inquiry of individuals, as experience there and elsewhere has abundantly shown.

Owing to various causes, which the reader can well imagine, and some of which will be adverted to in the course of this memoir, it will be proper to exercise more than usual caution in giving credence to relations in and about California, at least until the *auriferous excitement* shall have subsided.

During a conversation upon this subject—this peculiar state of things— it was remarked by one "not unknown to military fame," and whose official position for a year past has enabled him to see and hear much in California, "that it is safest to believe nothing you hear, and *doubt* half you see."*

---

* The following anecdote will perhaps aid in illustrating these views. A gentleman from the banks of the Rhine, whose good judgment and industry enabled him to amass a large fortune in mercantile operations in Mexico, was a fellow passenger in the steamer from Panama to San Blas. When he learned early in the spring of last year that uncoined gold could be purchased at a low price in California, he hurried thither with a large sum in silver; but the value of gold had advanced, and not being able to make a satisfactory exchange, he shipped his money to England via Panama. Among other things he remarked, "as soon as you reach San Francisco you will think every one is crazy; and without great caution, in three days you will be crazy yourself." He lost no time in getting off with his money.

Among the older residents, however, there are those who, if they have been afflicted with this moral fever, are entirely recovered from its effects, and are capable of furnishing reliable information relating to the resources of the country, especially those derived from the vegetable and animal kingdom.

California seems hitherto to have been regarded almost exclusively with reference to its capacity to produce gold; and though that subject received a due share of attention, yet other branches of industry were examined to as full an extent as circumstances and the brief stay of the writer in the country permitted.

Notwithstanding the *seeming* brilliancy of the golden prospects, the full development of those branches of industry embraced under the general head of agriculture, in connexion with the arts by which its products are elaborated, is far more important to the *permanent* prosperity of the country than its precious metals can ever become.

These considerations were sufficient to cause every suitable occasion to be availed of to collect facts and make observations in reference to agriculture, and those industrial pursuits that tend to augment its importance to mankind.

A small collection of specimens of minerals and fossils was made during the reconnoissance and shipped home by sea; a portion of which are intended to be deposited in the collection of the Smithsonian Institute when they shall arrive. The few fossils are from the tertiary (miocene) beds of the coast range, and will be carefully examined.

The want of accurate maps of the country rendered it difficult to give geological sections and sketches of the routes travelled, so that in constructing them the heights and distances could only be estimated; and as the thickness and boundaries of the various formations could not be determined by measurement, their localities are approximately written, without giving their boundaries, except in one or two instances where it could be done with an approach to accuracy. They are given with a view of enabling the reader to form an idea of the *prominent* geological features of the country, and are sufficiently accurate for that purpose.

I take great pleasure in acknowledging my obligations to John H. Alexander, esq., of Baltimore, for his efficient aid in reference to the maps and other drawings which accompany the memoir.

JANUARY, 1850.

I. GEOLOGY OF PART OF THE SIERRA NEVADA.

The great California mountain range, whose snow-capped peaks suggested the name of Sierra Nevada which has been applied to it by the Spaniards, constitutes so important a feature, in its relations to the geology as well as the physical geography of the country, that it seems proper to give a general outline of it before describing the geological character of the portions of California that will be the subject of this memoir.

The "*Great Basin*" is separated by this broad mountain ridge from the large *valley* of California, to which the names of its two principal rivers have been given, the Sacramento and San Joaquin, although it is in fact but a single valley. Between this valley and the Pacific ocean is the coast range, whose ridges and small valleys occupy a mean breadth of little more than thirty miles from the sea.

The geological equivalent of the Sierra Nevada to the northward is the Cascade mountains of Oregon, which extend nearly south, from latitude 49° north, between the meridians of 121° and 122° west of Greenwich, to about latitude 41° north. From thence, besides throwing out lateral spurs towards the coast range and the Sierra Nevada, it appears to sink away in a SSE. direction from Mount Shastl, and becomes merged in the western flank of the Sierra Nevada.

The Sierra first shows itself as a distinct ridge near the northern boundary of Oregon, in west longitude 120° 40', and extends in a direction nearly south to latitude 38° 30', from whence its course is southeasterly, but somewhat serpentine, until it unites with the coast range in about latitude 34° 30', as appears from the investigations of Fremont.

Southward of this point of union these two mountain ranges become one; which, in extending south southeasterly, in the form of a broad range of moderate elevation, constitute the peninsula of Lower California.

The Cascade range, according to Wilkes, attains an elevation of 5,000 to 6,000 feet throughout most of its length; but a number of volcanoes extend above its granite summits far into the regions of perpetual snow.

The observations of Fremont indicate that the ordinary altitude of the Sierra is greater than that of the Cascade mountains, whilst its volcanic peaks, also, are elevated above the inferior limits of perpetual snow.

These volcanoes, like those of Auvergne, have apparently forced their way upwards through granite.

In geological structure the Sierra Nevada much resembles the Andes; and the analogy holds, also, in being, like the Andes, one great ridge, instead of a chain of ridges, such as constitutes the Appalachian mountains east of the Mississippi.

The eastern flank of the Sierra has no great width, and is much confounded with the apparently irregular mountain masses of the " Great Basin."

Those portions of the western flank that came under my notice are from forty to fifty miles or more wide, with an average rise from the valley in a direct line to the *ordinary* summits not exceeding, probably, two degrees.

Fremont's map of 1848 shows a greater width of flank as it recedes from the latitude of 38° towards its northern and southern termini. My reconnoissance extended sufficiently far to the north to satisfy me of the correctness of his map, in that respect, above the latitude of 38°; and his opportunities south of that line were such as to enable him to give an excellent delineation of the mountain in that direction.

Having given this general sketch of these California mountains, I shall proceed to give a more detailed account of the geological structure of the portion of country embraced within the range of my reconnoissance, which was confined to the western flank of the Sierra Nevada, between the Yuba and Calaveras rivers and the portions of the Sacramento valley and coast range southwesterly therefrom.

Availing myself of the invitation of General Smith (who, with his suite, was about to make a rather rapid reconnoissance through a portion of the territory, accompanied by Commodore Jones and the Honorable T. B. King) I visited parts of the Sierra and traversed the Sacramento valley to an extent sufficient to form general conclusions in reference to the geological structure of those districts. Subsequently a portion of the country within the limits of the coast range was examined.

Instead of a journal of travels a sketch of the routes pursued will be given; not consecutively, but in the order that will be likely to contribute the best to a proper understanding of the subject to be treated of.

In making the tour within the flank of the Sierra Nevada the party entered the hills near the point where the waters of Bear river issue from them, and pursued a circuitous route about N. 20° E. to the Yuba river; from thence over an elevated and broken country nearly N. 65° E. for about twenty-five miles, nearly parallel to the Yuba, whose ravine was again reached. Upon leaving the Yuba the country preserved its uneven character; the general course was about SSE., inclining towards the axis of the Sierra and leading through a region elevated 3,000 to 5,000 feet above the level of the sea. We crossed successively Bear river and the North, Middle, and South Forks of the American river, and, a few miles south of the last named, moved westerly towards the "great valley," which was reached in a distance of about thirty-six miles, and the additional distance of about fourteen miles took us by Suter's fort to Sacramento City, near the mouth of the river.

Retracing 13 miles to the camp, we moved from thence in a SSE. direction over the valley, whose surface, to the point where the Cosumes emerges from the hills, was a dry prairie. From this point a tour was made through a portion of the mountain flank to and across the Mokelemy river. Near the Calaveras river the party divided: the General and part of his suite and Mr. King made a trip to the Stanislaus. Commodore Jones with the remainder left the mountain region and proceeded to Stockton, and the waning strength of my horse induced me to take the same course.

In addition to what has been already said in regard to the general features of the Sierra, it may now be stated that its western flanks, as far as observed, consist of a vast mass of metamorphic and hypogene rocks stretching from the Sacramento valley to the axis of the mountain. (See figures 1 to 6.) This mass of matter has an average slope from the valley upwards of 180 feet to the mile, thus giving a great rate of fall to the streams which rise in the vicinity of the snow peaks: these, aided by the decomposing energies of atmospheric agents, have excavated ravines of enormous depths, reaching along some branches of the American river at least 3,500 feet. Into these other ravines open with their innumerable tributaries, which, by intersecting the country in every direction, give it the appearance of a group of rounded and conical mountains.

The rocks of this region consist mainly of various slates and hypogene rocks, which have been protruded from beneath them.

We did not approach nearer the summit of the Sierra than about 20 miles, so that no account can be given of that space with certainty; but from what could be learned from the gold seekers, it is probable that the slates do not extend much further eastward before they are met by granitic aggregates, which, as before stated, exist at the summit.

The slate rocks contain numerous veins of quartz, which are without doubt repositories of gold, and probably of other metals and metallic ores. It is from the destruction of the ruins which formerly existed in the spaces now occupied by the numerous ravines of this region, that the gold which is found among the transported matters of the ravines was derived. These veins will be more particularly noticed in another place, in describing the geological structure of the region. Their great importance would have entitled them to careful examination, if such could have been made during our hasty reconnoissance. That the veins are very numerous, is manifest from the abundance of fragments of quartz often stretching in lines over the surface of the hills.

The low rounded hills that constitute the eastern border of the Sacramento valley near Bear river, where our party first struck them, are thickly covered with diluvial drift, consisting mainly of sand and loamy matters, mixed up with jasper, prase, basanite, and other silicious pebbles. At some points this heavy covering conceals the rocks of the hills nearest the valley. As the structure of the valley beneath the diluvial matters appeared to be a matter of practical importance independent of the scientific interest of the subject, these points were carefully observed.

Continuing on between and over these hills, whose elevation increased as we receded from the valley, the first rock met with was a trappean variety of a compact structure, very similar to that of basalt. This is in the sides of the ravines, the hills still having their diluvial mantles. The route pursued now gave a more rapid ascent, and the proportion of diluvium became lessened; and finally, when the elevated country near where we reached the Yuba was attained, there was no diluvium observed upon the hills: consequently the rocks became more and more visible.

Succeeding the trap rocks were slates; and, in fact, over the remainder of the 20 miles of our route, from Bear river to the Yuba, no other rocks were seen but the slates and the trap rocks that have found their way to the surface between them. The outcroppings of quartz veins were often observed; and, although we did not notice symptoms of gold digging until we reached the Yuba, yet it is more than probable that the metalliferous region embraced this district.

Near the Yuba the trap rocks contained epidote, actynolite, specular oxide of iron, and also small crystals of quartz.

During this part of our travel we had a fine view of the Buttes, through the remarkably clear atmosphere of this region. This striking feature of the Sacramento valley stands out in bold relief from the plain, and is surrounded thereby. It is evidently an extinct volcano, as appears from its mineral composition.

The point where we first struck the Yuba river is about 20 miles above its mouth. It is here a mountain torrent, running through a narrow chasm between the hills, which had been excavated to a great depth. This ravine is so narrow as to leave little space on the borders of the stream for alluvium. Huge boulders line its banks, except where the slate or trap rocks in situ are involved by the stream itself.

The large boulders serve to detain those of smaller size, as well as sand and gravel, which otherwise would be carried off by the resistless force of the periodical floods of this river.

It is by partially removing such deposites as these that gold is obtained —I say partially, because these large boulders being immovable under the *present* system, the digger is obliged to content himself with what he can get by scratching out from among these masses all in his power, and washing it for the metal it may contain.

We were informed at this spot that pieces of gold had been found upon the faces of the ravine at very elevated points. This, of course, may be expected to occur, as the crumbling of the *slate* rocks liberates the vein-stone (quartz) in which the gold was imprisoned; but on such steep hill-sides as these, the force of gravitation in no great length of time brings the gold to the bottom of the ravine.

It being impossible to travel up the ravine of the Yuba, a detour was made over the very elevated and rugged country through which it runs; and, after a journey over the hills southward of the river of about 25 miles, during which the sum of ascents and descents was probably not less than 15,000 feet, we reached the southern branch of the river at a point 12 or 14 miles from where we left the main river.

For about three-fourths of the way trap appeared to be the prevailing rock, with but little slate; then sienite seems to replace the trap, and continues (in connexion with slates in greater abundance) to the river.

The South Fork of the Yuba was reached by descending the face of its ravine where the declivity is exceedingly steep. This ravine has been cut down, not much if any short of 3,000 feet below the summits of the adjoining highlands. The strips of alluvium do not exceed in proportion those noticed below, and, like them, abound with huge boulders of hard hypogene rocks and indurated slates, filled in with smaller fragments of the same, as well as quartz, &c. Iron sand also abounds.

Leaving the Yuba, a very elevated and rugged region was traversed for some miles, consisting of sienites principally. The numerous ravines were of no great depth, owing in a great measure to the small amount of slates, which, being easily disintegrated, afford facilities for the formation of ravines. At about seven miles slates take the place of the sienites, but are again interrupted by an intrusive rock, which consists of a large area of serpentine, near the northern edge of which there is magnetic oxide of iron *apparently in abundance*. From thence trap and porphyry seemingly alternate with slates to Bear river, near which slate becomes the prevailing rock.

For several miles southeasterly from Bear river, the most common rocks are trap and porphyry, with, however, intervening slates. At one point a highly elevated mass of serpentine was seen, which was succeeded by porphyritic and trappean rocks, and afterwards by slates reaching midway to the North Fork of the American river. Another intrusion of trap and one of serpentine then succeed, after which slate becomes the predominating rock, with small intrusive masses of trap.

The North Fork of the American river, at the spot where the party crossed it, flows through a ravine which has been cut into the slate rocks to a depth of at least 3,000 feet. In approaching it, the descent is gentle at first, but increases towards the bottom to such a rate that one's own hands and feet are the only safe means of locomotion.

The strips of alluvium are very narrow here, as well as where Bear river was crossed. Large boulders of hard rocks also are sufficiently numerous to make the labors of the gold diggers, who line the borders of this stream, very severe.

Although it was thought the northern face of the ravine was very steep, the opposite one was much more so, and in fact it might have been deemed impracticable almost for horses or mules but for the horse paths in full view. Leading our horses, we surmounted by toiling up a few yards at a time, with the fierce rays of a Persian sun striking the arid and dusty surface at right-angles to it, and we had only here and there a stunted tree or shrub to rest under.

The slate rocks continued to the "Middle Fork," except at the end of the first five miles, where there was a development of serpentine for about two miles.

The ravine of the Middle Fork seemed not quite so deep as that of the "North Fork," and the approach to the river from the northward was found to have a more gentle descent. The bottom of the ravine also is wider, so as to permit larger deposites of alluvium, especially on the southern bank of the river, where the drift finds room to expand out to a width of one hundred yards. It is called *the Spanish bar*," and is presumed of one mile in length. The surface is but little elevated above ordinary freshets; and during great floods, which flow over it, the velocity of the waters is so far lessened, by reason of ample width, that the larger stony masses, which the resistless force of these floods is capable of rolling through the narrow passes, are arrested along the narrow strips of alluvium above the Spanish bar, and consequently its surface is covered mainly with gravel and sand. It is not to be supposed, however, that boulders of moderate size may not exist among the inferior portion of the drift, because such must have been deposited upon the bed and margins of the stream, as it formed this flat on its southern border by progressively excavating the base of the hills on the opposite side.

It was, at the time of our visit, a favorite spot for gold digging, and was worked by a large number of persons.

Leaving the river at the lower end of the Spanish bar, where the bottom of the ravine is again narrowed to its ordinary width of about one hundred yards, the route led up to the highlands, over an ascent of about 2,500 feet in four miles, then descended by a series of ravines two miles to the "Long Valley," and continued five or six miles therein. This is the longest valley, *properly so called*, that I saw in the more *elevated* portions of the flank of the Sierra, and it was said to be ten miles in length. Its breadth was estimated at four hundred to six hundred yards. There were smaller valleys in its vicinity, one of which was two or three miles long, and parallel, or nearly so, to the long valley into which it opens.

Such valleys are common where there is an extensive area of slate, as is the case in this district. So far as could be seen there were no other rocks between the Middle and South Fork of the American river. They abound in veins of quartz.

These slates are of unequal hardness in different strata, some of which disintegrate more rapidly than others. The minute state of division of the resulting earthy matters, enables the waters to carry them off without the aid of much descent in the beds of the rivulets that flow through these valleys in *winter ;* the harder slates being less acted upon from the dges between the valleys.

After leaving the "Long Valley," a region of more elevated ground, with many small valleys and ravines, was traversed for about seven miles, when we descended to the South Fork of the American river, by a declivity of moderate slope at the town of Coloma.

It was here that Captain Suter, in digging the tail race for a saw-mill, about 16 months previous to our visit, exposed to view the spangles of gold first seen in this part of California, and from which such extraordinary results have followed.

A thriving town has sprung up where, within so short a time, was a dreary wilderness.

There is a large area of alluvium along the river, and gold digging is carried on to a considerable extent in the vicinity of this place.

Our route was next to lead us out of the mountains in the direction of Suter's fort; but the river being, as usual, cañoned between mural precipices in so many places, we were obliged to take the wagon road (now so much travelled) which leads over the highlands.

The road winds its way up from the river, through ravines flanked by trappean rocks, for about five miles, the rate of ascent not being great.

At this point we descended into what seemed to be a basin-shaped depression some five or six miles in diameter. At both points, where we crossed the edges of this apparent basin, the rocks were trap.

The space within was filled with numerous knobs, or irregular cones of serpentine allied rocks. After leaving this basin, trap rocks continued for two miles, when we entered an elevated valley of irregular shape, and a few miles in diameter, called the Marsh Plain. A covering of detritus concealed most of the rocks in this valley, but it contained a few knobs composed of an aggregate of quartz, felspar, hornblende, and mica that might be taken for granite, which it much resembles in appearance, and from which it is only distinguished by the presence of hornblende. These rocks are seen at intervals for two miles westward of the Marsh Plain; and from thence to near Mormon island, slates, with trap interposed in many places, were the only rocks seen.

At Mormon island the South Fork passes between precipices, composed of an aggregate of felspar and a dark-colored mica (septenite) easy to be taken for hornblende; so that, at first sight, the rock somewhat resembles sienite. Slates border the river a short distance above and below this point, which is stated to be about four miles above where the South Fork is joined by the North Fork, (with which the Middle Fork had united a few miles further up.) This union of these principal branches takes place at the lowest point where gold has been worked, and from thence the name of American river holds to its confluence with the Sacramento river.

There is a tolerable width of alluvium on either side of the river at Mormon island, and on the south side there appeared to be a fair proportion of gold, if we may judge from the reported experience of the operatives then and there at work. The bed of drift on the northern side of the river is much higher than on the southern side. The diggers who essayed to work it stated that little gold could be obtained here, as at other places, until the whole depth of the drift is reached, which they found to be about sixteen feet.

A considerable amount of gold had already been procured along this portion of the river, and, under a belief that its bed must contain a very large amount, a company was engaged in attempting to divert the water into an artificial channel.

The parties were laboring with praiseworthy perseverance; but their plans and modes of working seemed to be so unartistic, as sadly to indicate the want of an engineer. Subsequently I was informed that although they did succeed in turning off a considerable proportion of the water, they failed to dry the bed of the river, and reaped a reward far from commensurate with their expenditure of labor and means.

Upon regaining the moderately elevated region southward of Mormon island, our westerly course led us almost imperceptibly into the Sacramento valley, in a distance of 7 or 8 miles.

The septenite rock before noticed at Mormon island was succeeded by slates with interposed trappean masses.

The highlands westward from Coloma reached an elevation of about 2,000 feet above the river, which at this place cannot be less than 3,000 feet above the sea. Therefore, in descending to the valley (whose surface is little raised above the tide-level in this part of it) we made a descent of about 5,000 feet. This was so gradually effected that it would scarcely have been suspected to be so great, unless the estimate be found upon the fall in the bed of the river, from Coloma downwards. The hills southward of Mormon island, which we passed, are about 1,000 feet above the ocean. They have rounded forms with intervening valleys, and would constitute a fine district but for the almost entire absence of water. The diluvial drift which begins here to cover them to some extent increases in quantity so as to conceal the rocks before the edge of the valley is attained. The geological structure is here precisely as where we entered the valley at Bear river, but on the line of travel from the latter the mountain flank rises much more rapidly.

During a few day's delay at an encampment on the American river, (a few miles below the point of junction of the North and South Forks,) a visit was made *via* Suter's fort to the new "canvass" city of Sacramento, which, although only three months old, contained more than one thousand inhabitants, who were driving a very brisk trade with those of the "diggings." They would not, however, be prevented by the pressing calls of business from giving the General, Commodore, and party, a sumptuous entertainment.

As the valley of the Sacramento will be considered in a separate chapter, no reference need be made to the portion travelled over on this occasion.

From the encampment on the American river our course was about SSE. to the edge of the valley, where the Cosumes river flows into it. At this point conglomerates and sandstones are to be seen beneath the more recent diluvial drift, with an almost imperceptible drift to the westward, which, in extending eastward, cover the older rocks.

About three miles above Dalen's mill a porous lava, or rather volcanic tufa, is exposed to view for a small extent about 200 feet above the level of the river near it.

I was informed that slate rocks were exposed in the ravine of the river, at a short distance from this spot, and that gold was to be had in the neighboring alluvium.

From thence a southeasterly course led over the hills covered by conglomerate and sandstone, resting upon a thick bed of indurated clay, and dipping westerly at an angle not much exceeding one degree; and although much of this more recent formation has been carried off, as is testified by the numerous irregular valleys and wide ravines that penetrate it in all directions, the metamorphic rocks beneath them do not appear even in the ravines until an elevation of 500 or 600 feet was attained—at a distance of five or six miles from where we left the river. Further on, the spaces covered by the sedimentary rocks become lessened by reason of the vast effects of floods which swept them principally during the more turbulent periods of the earth's career.

A few miles further, at an elevation of 1,000 feet, we no longer saw them. The slates became the only visible rock until, at twelve miles from the Cosumes, there is serpentine rock; and again, at two miles further, on the edge of the valley of Willow spring, there is serpentine associated with crystalline limestone, being the first I had seen. It is magnesia limestone, or dolomite, and does not appear to exist in great quantity. The country becomes more broken after passing out of this valley, and we traversed elevations more than 2,000 feet above the level of tide-water.

The ravine called "Arroyo Seco" furnishes much gold, and a large number of persons were engaged in digging for it. From thence an elevated slate region, intersected by numerous wide and deep ravines, extends to the Mokelemy, near which, however, the conglomerates and sandstones again appeared at a height of more than two thousand feet above tide.

Not far from Mokelemy are the ravines of Jackson creek and Suter's creek, (the latter almost dry,) which give employment to a large number of persons in digging for gold.

In fact, from the abundance of slate rocks between the Cosumes and the Mokelemy, with their innumerable ravines, there is a vast quantity of this metal within reach of *ordinary* digging.

This extensive development of slates embraces many varieties of that rock, differing materially in texture and hardness, owing to variations in composition, and in the temperatures they have been subjected to, as well as in the length of time they were excessively heated.

In some spots there is much hornblende mixed up with them, usually in the form of hornblende slate.

Descending from the highlands to the Mokelemy river, at a point some eight or ten miles from the borders of the Sacramento valley, there is a steep declivity where the river has scoured out a ravine to the depth of at least 2,000 feet.

Numbers of people were working out gold here when we reached the river, and it was stated the diggings were continued with activity above and below this point.

The route intended from here was across the river and over the highlands to the Stanislaus river; but upon learning its impracticability for the baggage team, we countermarched to the highlands by retracing our route for four miles. From thence about twenty miles of a circuitous route, over a very elevated and broken country, led us to the river again, at no great distance below where we left it.

Soon after leaving our former trail large masses of porphyritic and trap rocks were passed over for about two miles.

They continued some miles further, apparently alternating with hornblende and talcose slates, when the last named became the only rock. We next traversed for seven miles a deep basin-shaped depression, and found slates to bound it at the points over which we passed on entering and leaving it; within this area were numerous small hills, or rather plateaux, composed of conglomerate, lying horizontally, or, perhaps, dipping very slightly to the westward.

This conglomerate is highly ferruginous, and it would have been interesting to have investigated it for the purpose of ascertaining whether it be not a more recent deposite than the great beds before spoken of, and which are so abundantly developed in California; but there was no opportunity to do so. Slates apparently surrounded the basing, and continued to the Mokelemy river.

Gold digging is prosecuted successfully at this place, where the ravine is wider than where we approached it above, and the alluvium consequently more abundant. After crossing the river at a ford filled with boulders of *all* sizes, we passed along an alluvial flat 200 yards wide for more than one mile, to the lower end of it, where the ravine is again contracted to a narrow defile, forcing us up a steep hill more than 1,000 feet above the river.

Moving southeasterly from this point, we passed a series of alternations of hard and soft slates and chlorite slate, and in several places indurated talcose slates, which, besides veins of quartz, contained a notable proportion disseminated through its mass. These rocks were traversed for 5 or 6 miles, where they were interrupted by a hill of volcanic tufa (much of whose surface was covered with boulders and pebbles of various volcanic rocks) for two miles, when slates, similar to those last noticed, again appeared and continued for 7 miles with an interruption, over one short distance, of an intrusive mass of serpentine. Much of the surface within the last 5 of the 7 miles was covered with thick beds of conglomerate, or more properly breccia, because the volcanic pebbles which compose it are scarcely water-worn. Boulders, some of which are nearly a ton in weight, and pebbles of volcanic rocks, are strewed over the surface in this vicinity. Near our camping ground at the Double spring there is a rugged-looking lava, and also pitchstone, beautifully colored and variegated; besides silicified wood.

At this point the General and most of his suite, with Mr. King, left the waggon train for a trip to the Stanislaus river, which occupied a few days; but as my horse gave indications of not being able to encounter that task, I accompanied Commodore Jones and others to Stockton.

The volcanic rocks continued for one mile, and then slates about 3 miles, which again reappeared after a short interval of serpentine. Our SSE. course was now changed to about ESE., which rapidly took us out of the mountains, about 6 or 7 miles from the Double spring, and about 4 from the Calaveras, where the slates again sink away under the conglomerate, which in its turn passes under the more recent sedimentary deposites of the valley of the San Joaquin.

At the Double spring the elevation is not less than 1,200 feet above tide; but within a few miles there are hills somewhat higher; after which they are less elevated and more rounded, and are separated by wider valleys. This beautiful rolling country finally sinks into the "Great Valley."

The conglomerates which cover many of the hills between the Cosumes and Calaveras are found at the height of more than 2,000 feet, and consist of pebbles, with a cement more or less ferruginous. They are in thick strata, with intervening beds of sandstone of lesser thickness; and beneath them are heavy deposites of indurated clay, varying in color from dull gray to white.

(There was no opportunity to determine the thickness with an approach to accuracy, but it is not probable the entire formation much exceeds 200 feet.)

It was very desirable to procure the fossils of this formation, but none were met with from which its period can be precisely determined.

Continuing the journey, the valley of the Calaveras came into view, as we gradually approached and afterwards crossed it, the river itself being nearly dry. From thence we proceeded to Stockton, situated on a creek, or "slough," which runs 3 miles to the San Joaquin, a few miles south of the Calaveras.

In company with Commodore Jones, I travelled from Stockton, about south 13 miles, to Bonsal's ferry, on the San Joaquin, and after crossing there, pursued a westerly course across the valley 18 miles, to the foot of the eastern hills of the coast range. Our route over the valley was upon the main road leading to San Jose and Monterey, which was followed about 20 miles further in a course nearly due west, among the ravines and valleys of the same range.

In the following chapter will be given a sketch of the geological features of the limited portions of the coast range district which were traversed.

II. GEOLOGY OF THE COAST RANGE.

Upon leaving the valley just now spoken of, our road led up a crooked ravine in sandstone. At the mouth of this ravine this sandstone is seen, at first with a very slight dip to the east, running under the diluvium and the newer deposites of the valley, as was the case on its eastern side at the Cosumes, Calaveras, &c. The rate of dip progressively increases in pursuing the ravine to its head, a distance upon our crooked route of eight miles. At this point a low narrow saddle separates this from another still more circuitous ravine, running westerly. On the eastern side and near this summit, the sandstone beds attain a dip of perhaps 30°; at the anticlinal axis their dip is suddenly changed to the west with a high angle for two miles, when it as suddenly changes again to the east.

A series of anticlinal and synclinal axes then succeed, producing a wave-like stratification for three miles; at which point, on emerging from the ravine, it is observed passing under a valley which we crossed for four miles. Beyond this valley several low hills were passed, and another valley reached containing Livermore's ranch. The low hills were composed of sandstone, but their stratification could not be made out. Specimens of large fossil oyster shells were observed here, and Mr. Livermore informed us that he had obtained large bones, which he supposed to be those of a whale, and he was formerly a whaler himself. They occurred in a rather soft marl, and soon crumbled to powder upon being exposed to the air. After leaving Livermore's valley, which opens into Armidor's valley, we crossed the latter about six miles to the foot of the high mountains east of it. Those we passed *through* do not exceed in elevation 1,000 feet.

Near the foot of the mountain at Armidor's a sandstone rock is seen, containing many crumbling fossil shells, only one of which was sufficiently firm for a specimen, and was secured.

The strata here dip to the eastward.

Since leaving the Sacramento valley, the route so far formed about a right-angle with the mountain range; but from Armidor's ranch we followed the narrow valley parallel to the strike of the strata (which is a little to the west of north) for 13 miles. It gradually narrows to this point, and is terminated by several low hills; and after passing a mile and a half over them, another valley is reached and followed about four miles, where is a depression or saddle in the hills on the east, over which the road led into the Martinez valley, which we followed to Martinez, a newly-laid-out city, on the straits of Carquinos, opposite Benicia.

Sandstone, &c., continued to make up the hills from Armidor's, dipping at a high angle on either side of the valleys to the westward. The larger fossil oyster shell is also found near Martinez.

The straits of Carquinos, through which Treusum bay discharges its waters into the bay of San Pablo, are from one to three miles wide, and extend through a chasm on this portion of the coast range nearly at a right angle to the direction of these mountains.

The identical strata of one side is seen on the line of strike on the other; a proof that this outlet was opened at farthest since the very recent era of the earth's history, when these rocks of the eocene or miocene periods were formed. Sandstone is the principal rock, with a few intervening strata of conglomerate of moderate thickness, and beds of indurated clay, (here of a bluish gray color.) The pebbles of the conglomerate are identical in composition with those observed on the hills eastward from the San Joaquin, but of *much smaller* size.

Near the eastern entrance of the straits there is exposed, just at the water's edge at the lowest tides, irregular seams filled with matter of igneous origin, apparently trachytic, which must have been injected from below. The effects of heat are apparent upon the sedimentary rocks adjoining.

The most eastern ridge of the coast range, in stretching from the southward towards Suisun bay, appear to lessen in altitude until it unites with Monte Diablo, whose northern base nearly reaches the shores of the bay. This mountain forms a prominent feature in the landscape, and its cloven summit is seen from far and wide at points within the Sierra Nevada, as well as in the intervening Sacramento valley.

An opportunity to examine this mountain, as was intended, did not occur, but its appearance from points near its base indicates volcanic origin. Constant springs rise near the summit, which is about 3,800 feet above the sea.

Leaving the northern side of the straits at Benicia, we find there has been a still greater proportion of the hills removed in the space now occupied by the northwestern part of Suisun bay. For about three miles a bold escarpment exposes the strata, and the undermining action of the waters has aided in forming a talus of blocks of sandstone of considerable extent. The stone is of an excellent quality for architectural purposes. The whole talus may be considered an inexhaustible supply of building-stone, ready quarried, and deposited on the very edge of water so deep that large vessels may lie in contact with the rocks, if *they wish*. There are near the straits some thin beds of indurated clay, with a fissile structure, containing nodules of the common lethoidal carbonate of iron sufficiently good (if there be an ample supply) for tne manufacture of iron; useless, however, here, because of the *total* absence of coal and the sparse supply of wood.

Associated with these beds are others of an impure brimstone, about one foot in thickness, separated from each other by seams of clay and argillaceous sandstones. The external appearance of this limestone induces a belief that it would make an excellent hydraulic cement, if properly treated. If so, it will hereafter attain importance in connexion with the various constructions that must be made upon the shores of these bays, public as well as private, and should be carefully examined.

From the northwestern terminus of the escarpment, the shore of the bay tends more to the eastward and from the hills. A deposite of alluvium then commences, increasing in width towards the northeast, augmenting daily by deposites upon the extensive tule marshes which connect it with the bay. The road travelled gradually diverges from the ridges, upon which the sandstone and its associates continue for about 7 miles further, where, at the distance of several miles from the summits, small boulders of trap rocks were met with opposite to where the rounded summits became supplanted by those of a more rugged character, indicating the protrusion of hypogene rocks, which, upon again approaching the hills further north, was found to be the fact. Trap-rocks appeared and continued to prevail on the western hills up to the point where we left them by moving in a northeasterly direction.

The height of this ridge is greater as it extends to the NNW. between Suisun and Nappa valleys. Continuing for seven or eight miles in the Suisun valley, the easternmost range of the coast range was passed at a depression therein, no great distance north of where it falls off gradually into the low grounds bordering the northern limits of Suisun bay. On the western slope of this ridge a diluvial covering concealed the rocks. The central portions are decidedly of igneous origin, and the eastern flank is covered with sandstone, whose strata, with a decreasing dip to the ESE., pass under the diluvium of the Sacramento valley. This ridge also attains a greater elevation farther to northward.

Having completed this brief sketch of the structure and minerals of the eastern portion of the coast range, on lines north and south of the bays of San Pablo and Suisun, it may be remarked, that it would have been very desirable to have continued them both on to the shores of the ocean, even in the hasty manner in which the reconnoissance was executed, if it had been convenient to have effected it.

Observations at San Francisco supply one more of the missing links, although a little out of the line; and the examination of a considerable area, inland from Bodega bay, supplies all the remainder, except 12 or 15 miles of the northern line, including Sonoma and Nappa valleys.

The northern side of Bodega bay is separated from the ocean by an elevated narrow tongue of land stretching to the southward about three miles, and terminating in Bodega Point. This strip of land consists of beds of conglomerate, sand, and clay, reposing upon massive septenite, much intersected by fissures filled with flesh-colored felspar. These fissures or veins are confused, extended in every direction, having been injected whilst in a state of fusion, into the previously consolidated septenite. Appearances indicate that this strip of land formerly extended to a greater breadth seaward, and that it has been narrowed by the encroachment of the waters of the Pacific ocean, whose waves have surged against the rocks from the date of their elevation.

Huge masses of rock have fallen, and now lie at the foot of the cliffs, having been undermined by the excavating power of the waters.

The septenite rock, at the face of the cliffs, attains an altitude of perhaps 100 feet; but, on the eastern side, it reaches only eight or ten feet above the waters of the bay. Upon this, dipping easterly, and extending nearly down to the level of the water in the bay, is the conglomerate, whose component pebbles are very feebly cemented together. There then succeed beds of sand, with thin strata of clay intervening. The grains in the inferior portions of the sand are slightly held together by a ferruginous cement, and might be deemed incipient sandstone. These beds form an aggregate thickness of about 60 feet on the east, and thin off very much next the sea. Among the smaller beds of clay and sand nearest the conglomerate there are irregular masses of earthy hydrated peroxide of iron, and in the inferior portion of the conglomerate are many pieces of wood, not in the least mineralized, but in that stage of decay familiarly called *doted*.

The wood is exeogenous, and will be more particularly examined upon the arrival of the specimens. No other organic fossils were met with to assist in fixing the relative age of this deposite. It is more than probable it was formed and raised from the deep, long after the upheaval of those hitherto noticed upon the coast range.

It is by no means certain that it may not prove to be an upper tertiary fresh-water formation; one separated from the sea by a bulwark, of which all that now remains northward of Point Tomales is the narrow strip of land before noticed, and a small island of septenite, half a mile distant, in the prolonged line of its direction.

On the eastern side of the bay, extending one mile and a half from the small strip of recent sedimentary matter skirting it on that side, is a curved sand-bar, from 20 to 40 yards wide, whose materials are thrown up by the ceaseless action of the surf, aided by the winds blowing the sands to a still higher level, where they are retained by plants which have already taken root.

This bar leaves but a narrow passage between its terminus and Bodega Point, and thus secures the inner harbor as a safe one in all weathers. Both the destroying and forming effects of the sea are finely illustrated in this vicinity.

The first ridge to be surmounted in going westward from the bay is about three and a half miles wide, and is intersected by numerous ravines cutting deeply into it in every direction; but the facility with which disintegration takes place in these rocks is such, that there are few places where they can be seen, because the surface is almost everywhere rounded off and covered by detritus.

The ridge attains an altitude of about 800 feet, and appears to consist of trachytic rocks, covered for the most part with sandstones, &c., the equivalents of the formations before noticed on the eastern ridges of the coast range.

Westward of this ridge is an irregularly-shaped valley, upwards of half a mile wide, and four or five miles long, where is situated the homestead of the hospitable Captain Stephen Smith.

The northwestern end of this small vale is abruptly terminated by steep acclivities, whose rocks mainly consist of several varieties of compact quartz, such as chert, jasper, &c. There is every shade of white and gray to nearly black, as well as yellow, red, and brown, each color continuous in separate layers. It is evident they were originally deposited in strata; and, besides having been changed in character by heat, were contorted and twisted in a remarkable manner anterior to, or during the period of their elevation. Still further to the northward there exists a large area of mountainous country, of rugged character, attaining an elevation of 2,000 feet, or 3,000 feet, within five or six miles, and still higher further north. The rocks are hypogene and matamorphic, of many kinds, which have been twisted about and mixed together in the most confused manner. It is very uncommon to find so many kinds of rock in such close proximity. Within a space of a couple of hundred yards in diameter (and about two miles north of Captain Smith's residence) there exist gneiss, mica-slate, indurated talcose slate, hornblende slate, and serpentine, the last containing chroniferous iron. Crystals of actinolite and black hornblende were found in the hornblende slate. In the vicinity is also steatite, of the fine variety called French chalk. Trachytes seem to prevail between the above and the cherty rocks before noticed, and they are covered to some extent by sandstones, which spread over portions of the hills on either side of the valley, as well as in the valley itself. In addition to the extensive development of the quartzose rocks in the northwestern part of the valley, there are large isolated masses of them protruding above its surface (but lessening in size and number) towards the southeast.

Amongst these there is in one spot a *small* area of *very* course conglomerate, which has been tilted out of a former position. There was no other rock resembling it seen in this part of the country, save that at Bodega bay, which was less firmly cemented.

Some imperfect marine shells exist in the fine-grained sandstone, of which specimens were obtained.

In the more elevated country in the northeast is a larger extent of talcose and argillite slate, with veins of quartz in proximity to hypogene rocks, which also contain quartz veins. There is little doubt that a careful investigation would demonstrate the existence of gold and other metals, or their ores, in these veins.

In proceeding eastward from Smith's valley, rounded hills increasing in height are found for about four miles, when a ridge of some one thousand five hundred feet high intervenes. Trachytes are the only rocks visible until the summit is passed, at two miles further, when talcose slates are seen for a short distance.

From thence the country is made of minor ridges, decreasing in altitude to Petaloma valley. Their geological structure is much hidden by detritus, giving them the smooth rounded form so common in this region.

Trachytic rocks occasionally are seen in situ, and probably constitute the main bulk of these hills, covered, perhaps, in some places with the sedimentary rocks before noticed.

Eastward of Petaloma valley is an elevated ridge, separating it from Sonoma valley. The western slope of this ridge is covered by sandstones dipping westward.

I did not examine the ridges between Petaloma and Snisun valleys. It was, however, manifest in sailing along the bays of San Francisco and San Pablo, that igneous or hypogene rocks are extensively developed on the elevated mountains between these bays and the ocean. From the bay of San Francisco along the seacoast to Bodega bay, there are bold bluffs apparently of similar rocks.

Westward of San Francisco we find slates, much twisted and confused by the trachytes, which have forced their way through, and in part cover them, as is seen in the escarpment immediately west of the city. On the southeast, at Rincon Point, there is a confused mass of trachytic rocks much mixed up with quartz, disseminated as well as filling small seams. There were also small fissures filled with calcareous spar. The slates here, too, furnish evidence of having been much disturbed during the period when the igneous rocks were protruded through them. Carbonate of copper is sparingly seen at this locality.

Circumstances prevented my making a reconnoissance of the region between San Francisco bay and San Diego, as had been intended.

Upon my return home we remained two days at the latter port, and I had an opportunity to observe that sedimentary rocks prevailed on the hills near the sea, and that there was an extensive diffusion of diluvial drift.

It had been reiterated over and over again in letters, newspapers, and in other ways, that there was, a few miles north of this port, near the seashore, a coal formation capable of furnishing ample supplies of the "best of coal for steamers" and other purposes. These beds prove to be layers of bitumen an inch or two thick, alternating with thin strata of small gravel and sand.

III. GEOLOGICAL STRUCTURE OF SACRAMENTO VALLEY.

In giving an account of this great valley Frémont calls it a single valley—"a single geographical formation near five hundred miles long." In addition, it may be said, its geological structure also proves the correctness of considering it a single valley. It is in fact a long prairie, occupying the space between the flanks of the Sierra Nevada and those of the coast range, closed in on the north by the terminal spurs of the Cascade mountains, and on the south by the junction of the coast range with the Sierra Nevada.

Its greatest width is little less than sixty miles; but it maintains a mean width of nearly fifty miles throughout almost its entire length. Near its northern limits the Sacramento river, leaving the valley in which it takes rise, flows through a deep chasm in a spur from the Cascade mountains, and runs through the central portions of this valley until its waters mingle with those of the San Joaquin, which rises in the mountains at the southeastern end of the valley. This meeting of the waters takes place in the delta at the head of Suisun bay. From this, the lowest part of the valley, the surface rises towards its northern and southern termini, almost imperceptibly at first, but at an increasing rate towards the sources of the rivers. In the course of the reconnoissance, besides crossing this valley twice, favorable opportunities were availed of for investigating its geological structure on its eastern edge between the American and Calaveras rivers.

It will be recollected that frequent mention was made of the sedimentary rocks which cover a large area of the hills of the western flank of the Sierra, especially between the Cosumes and Calaveras, and that these strata uniformly dipped very gently to the westward, and extended under the diluvial drift of the eastern side of the valley. In crossing the valley westward from the Calaveras, *this formation is again seen*, emerging with a moderate dip from under the more recent deposites on the westerly edge of the valley, and covering the eastern ridges of the coast range.

About twenty miles northward from Suisun bay it was again noticed under precisely the same circumstances.

No organic fossils were obtained on the eastern side; but a careful examination of the position, structure, and components of these strata leads irresistibly to the conclusion that they are identical and continuous. It is true that the *conglomerates* abound more on the east, and are composed of larger pebbles; but that fact only proves those on the eastern side of the valley to be nearer the rocks from whose debris these conglomerates were formed. As these strata were not seen upon the edges of the valley near Bear river and the American river, respectively, they cannot there reach so high a level as near the Mokelemy; and it is not likely they rise much above the level of the surface of the valley, or they would have been seen upon the sides of the ravines in the diluvial rolling ground near Bear river. It seems, therefore, that near the Mokelemy the upheavals of these beds continued to a period long after it ceased further north.

Before I visited this region there were two matters of interest that in an industrial point of view claimed especial attention. The first was, whether a coal formation might not exist under the great valley; and the second, whether its geological structure might not indicate means of supplying water for agricultural and other purposes.

The first query is answered by the fact of finding the comparatively recent strata of a formation certainly not older than the eocene or miocene eras, resting immediately upon the metamorphic and hypogene rocks of ancient origin—the remaining members of the *secondary*, with all the sedimentary rocks of *older date*, being wholly wanting, and *with them* the coal *formation* which belongs to the lowest of the secondary series. A coal formation under the valley is, therefore, out of the question, unless *deeply* seated and entirely covered, edges and all, by the sedimentary strata before noticed.

The soil of this valley, except on the lower portions of it near the Tulé swamp and on the borders of the very few streams which traverse it, is extremely arid during two-thirds of the year. It is common to travel long distances without seeing a trace of water. In going from Putu creek to Cash creek, fifteen miles, we traversed in July a parched plain entirely destitute of even moisture.

Unless ample supplies of water can be elevated above the surface of this valley, it can never attain much importance for the purposes of agriculture. It is true that the finest crops of wheat may be raised on much of the surface; yet summer crops, except barley and oats, are altogether impracticable, unless it be on the margins of the few streams.

It was therefore, in my opinion, very important to ascertain whether the geological structure beneath would warrant the expectation of bringing water above the surface by means of artesian wells, and I do not hesitate to say that it does most strongly.

Experience has demonstrated the almost certainty of procuring water by this means in all valleys resting upon sedimentary formations, having a basin-shaped stratification, and composed of beds, some of which are not too compact to admit the passage of water through or between them, while others are sufficiently compact to prevent the escape of the water filtering through such permeable strata.

Now these conditions are completely fulfilled in the valley of the Sacramento. We have sufficient evidence that the beds of conglomerate, sandstone, and clay which are seen to dip under the valley on its eastern side, are the same which rise out from beneath it on the west; and if the basin of London water will flow to the surface through wells of one hundred to fifty feet deep, in how much greater quantity may it be expected to flow beneath the clay-beds under this valley.

It is reasonable to suppose, in view of the facts and inferences, that this valley, so much denounced by disappointed seekers after gold, will, if the means above indicated for procuring supplies of water be generally resorted to, become exceedingly productive.

As such a large portion of it is public lands, the government should be foremost in adopting measures to improve their value by the means proposed.

IV. REVIEW OF THE GEOLOGICAL CHANGES IN CALIFORNIA.

Having now given sketches of the geology of the portions of California that came under my notice, sufficient, perhaps, to convey a general idea of their mineral structure, it may be proper, before considering the industrial applications to be hereinafter indicated, to give some general views, and to revert to the theoretical branch of the subject.

As before stated (upon the authority of Colonel Frémont,) granite is the prevailing rock near the summit of the Sierra, and it may be presumed that this or other igneous rocks reach some distance down the flank of the mountain.

The slates generally extended, as I was informed, but a few miles easterly of our line of travel. From thence to the valley the rocks may be divided into—1st, metamorphic, consisting of those of sedimentary origin, such as slates, but subsequently *altered* by the effects of heat; 2d, hypogene or rocks formed far beneath the surface, and brought into their present state by heat of sufficient intensity to have fused them before or during their elevation. Some of these are crystalline—as granite, septenite, &c.; others, as some of the trap-rocks, possess a close, compact structure, without any appearance of crystallization.

There was by no means so great a variety of metamorphic rocks observed upon the western flank of the Sierra as might have been expected. North of the American river no others were seen but varieties of talcose slates and argillites. Between the Cosumes and the Calaveras there were, in addition, hornblende and hard silicious talcose slates. This latter region also furnished testimony to a greater intensity of the action of heat upon the metamorphic rocks in the greater abundance of indurated and hornblende slates.

These slates, originally formed of sedimentary matters under water, have been hardened and otherwise changed in structure by the heat from the liquid masses below, which, in addition to elevating them, has in many instances caused them to burst through and find their way to the surface. The extensive region throughout which these slates abound shows that they once occupied a large area, which has been materially lessened by the intrusion of masses of igneous rocks in innumerable places. The intrusive rocks in this region are trap, septenite, porphyry, serpentine, &c.

The whole have been elevated to their present position evidently by forces from beneath; but the nature and immediate cause of action of those forces being an unsettled point in science, it would be out of place to discuss it here.

The region between Feather river and the American river bears evidence of having become quiescent, or nearly so, long before the *comparatively* recent period, when the country south of the Cosumés was in a state of disturbance, produced by a most energetic activity of those internal forces which have brought to the surface lava, tufa, &c. These volcanic masses do not seem to have been produced during the era within which the intrusive rocks between the American and Feather rivers were formed. Still *further* north volcanic action continued also to a later period, as well as further southeasterly. On the Stanislaus, General Smith informed me he saw vast developments of basalt. The upheaval was continued towards the northern and southern ends of the valley, after the portion between the Yuba and Cosumes became quiet or nearly so; and this is probably the main reason why the Sacramento valley is so much higher near its termini than where its united waters forced their way through the ridges of the coast range to the sea.

We have evidence in the existence of sedimentary rocks near the Mokelemy river, that they have been elevated 2,000 feet at least since their formation, which is certainly not *anterior* to the eocene period.

On the contrary, the absence of sedimentary deposites, older than diluvial drift, between the American and Feather rivers, proves the elevation of that region before the era above referred to.

To avoid repetition, the subject of *veins*, before alluded to, will be considered when treating of the gold region.

The elevation of the parts of the coast range I visited seems to have been contemporaneous with, or more recent than, that of the region about the Mokelemy, as may be supposed from the fact of sandstone covering them at a considerable altitude. The prevailing hypogene rocks are trachytes, which in some localities have forced their way through the sedimentary strata, leaving evidence of great disturbance, as near San Francisco and NE. of Bodega bay. In many places, however, the strata, instead of being broken through, are elevated in a series of parallel ridges— a feature which distinguishes the coast range from the single broad ridge constituting the Sierra Nevada—at least in the central parts of their course.

Besides the trachytes, there are also trap rocks, septenite, and serpentine. The latter has been protruded in numerous isolated masses, each occupying a small area.

Northward of the country I visited, and between the Russian river and the Laguna, Lieutenant Revere, U. S. N., mentions the occurrence of obsidian, usually an evidence of volcanic action of comparatively modern date. He also travelled over "*basaltic fragments*" in the same region.

At Benicia I found one piece of obsidian; and as there exists no evidence of its having been transported during floods, it may be presumed to have been brought from the locality above noticed by human agency.

Still further north, it is affirmed, lofty volcanic peaks shoot up from the coast range. Southwestward of San Francisco, as was before stated, the mountains also attain a considerable altitude, and furnish ample testimony of volcanic action; but not having traversed the country in that direction, I shall say little in reference to it.

This region contains the mines of native sulphuret of mercury so often referred to in connexion with California; and at no great distance therefrom there is an ore containing silver. Specimens of copper were shown to me, said to have been procured in the coast range northward of Monterey.

I was unable to procure reliable information of the geological position of those metals.

One of the proprietors of the "quicksilver mine," as it is termed, in reply to an inquiry, stated that it occurred in granite, which is altogether improbable. It is, however, not uncommon for other rocks to be taken for granite by persons not familiar with their characters.

The reader may have noticed, whilst sketching the geology of the gold region upon the flanks of the Sierra Nevada, it was observed that slates appeared to be the prevailing rocks there, as in all other gold-producing districts of any importance hitherto known. These slate rocks, it is stated by Professor Dana, extend northwardly into Oregon, where it may also be expected the gold will be found. Circumstances favorable to the occurrence of gold exist in that Territory. These may be briefly stated to be, the existence of veins of quartz in slates, in the vicinity of, or penetrated by, trap, porphyry, or other rocks of igneous origin.

It is not, therefore, surprising that gold should have been found in the soil of San Francisco, because, as has been already stated, the slate rock is there exposed, with small veins of quartz, in which the gold *may be* found, upon sufficient examination.

Whether it will ever pay for the working, is another question. There is no doubt but the little found there in the soil, and which created a "sensation" during the early part of last summer, came from its former positions in the veins of quartz in the disintegrated slates which formed a part of those now existing in the adjoining hills, much distorted by intrusive rocks.

It has long been known that gold occurred in the transported deposites of ravines on the western flanks of the coast range, (near San Fernando,) and in the vicinity of the point where it unites with the Sierra Nevada.

The scarcity of water for washing has hitherto prevented the profitable working of these "placers." There is no doubt but the slates heretofore noticed in other parts of the coast range extend in this direction, and probably continue to the southern terminus of Lower California.

North of San Francisco bay, slates were noted, stretching northward; and eight or ten miles northeast from Bodega bay, they are extensively developed, under circumstances very favorable to the prospect of finding gold, upon proper examination. Still further north, upon Trinity river, in latitude 41° 30', I learned, on the eve of leaving California, that gold was also known to exist, and a company was about to proceed there.

There were many reports of the existence of cinnabar; but the several specimens shown to me from such reported localities, which, with others, I examined personally, were rocks of various kinds, colored red with peroxide of iron. In fact, most stones thus colored are suspected in California to be this valuable ore, which gives rise to frequent false reports in reference thereto.

Lead ore—probably the sulphuret, or galena—occurs northward of Sonoma.

The principal rocks which have been met with having been noticed in the preceding pages, some reference may be made to the effects that have been produced upon them by atmospheric and other causes, which have so materially affected the character of the surface of this country.

All rocks are more or less dissolved or disintegrated when within reach of the oxygen and carbonic acid of the atmosphere, and water, aided by heat. Some, it is true, appear to be nearly indestructible, and are practically so when kept dry, as appears from the integrity of certain sienites in the dry climate of Egypt, of which some ancient monuments still exist; but even these would suffer, if slightly covered with earth, and kept wet. Many of the rocks in California are of the easily-destructible kinds; and it is mainly owing to this fact, that ravines of denudation are so exceedingly numerous, not only on the flanks of the Sierra Nevada, but throughout the region of the coast range, where both kinds of valleys exist.

In the last-named district, there are numerous valleys, before referred to, which, like the Sacramento, are valleys of elevation, produced by upheavals on their borders. Besides these, there are also valleys and ravines of denudation as numerous, resulting from the removal by water of the matter which occupied the spaces between their borders. These *valleys of elevation* are parallel to the ridges, or the range; and, as they form an important feature in reference to the agriculture of California, they will be again reverted to, under that head.

On the flanks of the Sierra Nevada, there were few, if any "valleys of elevation." "The Long Valley," before noticed, and others, of a smaller size, observed among the slates, may possibly be such, because their surfaces, in the direction of their greatest length, had but little descent, so that the beds of their small ravines of drainage have not sufficient fall to produce much excavation during winter, when water flows through, but which I usually found, to the discomfort of my horse and self, perfectly dry in midsummer. Even these valleys, though, may have been produced by the gradual disintegration of slates of a softer texture than the slates or other rocks in the adjoining grounds. A careful survey might determine this point.

In traversing this region of numberless hills, it is easy to see that the deep chasms and ravines between them were mainly produced by the disintegration of the rocks, and the excavating and transporting power of water.

Slate rocks, such as abound in this district, are among those most readily acted upon. Even in summer they have been elsewhere observed, incessantly exfoliating and producing the most minute fragments; but, in the climate of this mountain region, such effects are produced most rapidly by the freezing of absorbed water. The oxygen of the air also aids in their destruction, as well as of most other rocks, from the facility with which it acts, by separating their protoxide of iron and converting it into peroxide. Carbonic acid, when aided by water, is also unceasingly separating and combining with thin lime, magnesia, and alkalies. The effects of this acid are, although apparently slow, more potent than a casual observer would imagine.

I have observed myself that letters more than one-fourth of an inch deep were obliterated in 21 years from the surface of a magnesian limestone in Baltimore county, Maryland.

Even with the floods of the present day there is no difficulty, if sufficient time be admitted, in accounting for these wholesale "deep cuts;" but greater floods occurred in former days. The boulders scattered in the region of the Mokelemy, even on the highest grounds, prove that the vast floods which prevailed during a former era of the earth's career, and which bore icebergs from the northward over Europe, Asia, and the northeastern portion of North America, also carried them over parts of California. The great area of diluvial drift, and the elevation at which it is found, furnish evidence to confirm the theory of these floods, so fully established by the labors of many distinguished geologists. It was during this long continued diluvian era that denudations were most rapidly effected. It was then that the large valleys, before noticed, south of the American river, were mainly formed, by the removal of perhaps one-half the area of the conglomerates, sandstone, &c., which once covered the entire surface of the flanks of the mountain (at least between the Cosumes and the Calaveras) to an extent of not less than 20 miles eastward from the valley. And, more than this, it scoured out innumerable ravines among the slates and other soft rocks beneath them, which were thus again exposed to the light of day.

It was during this period also that most of the boulders were driven furiously along the ravines and lodged here and there, wherever the lessened velocity would permit them to remain, as stumbling blocks to the gold-seekers of our day. During all this turmoil, the gold imprisoned in the veins of the rocks thus destroyed was liberated and borne by virtue of its great specific weight to the bottoms of the ravines, where the force of the current, from the rapid fall and great depth of the torrents, was sufficient to leave nothing else of less comparative weight behind.

As these currents decreased in force, the heavier stony fragments began to be deposited and mix with and cover the auriferous and other metallic substance; but the gravel and sand continued on, and formed immense beds now seen upon the rolling country at the foot of the mountains, whilst much of the finer matters were deposited in the Great Valley, and even carried into the ocean itself, through the openings formed in the coast range probably during this era.

As the foliated structure of slates is a subject of some scientific interest, a reference to it may be expected in connexion with a region wherein this class of rocks is so extensively developed. It was *uniformly* observed that the lines of cleavage of the slates in the Sierra Nevada bore a course about *north by west*. These lines of cleavage are usually taken for those of stratification by persons unaccustomed to such investigation; but they are very different. It is frequently difficult to determine the stratification, even where extensive excavations have been made into such slate rocks. No opportunities were presented for investigation of those we saw in the Sierra Nevada. It is quite probable that the derangements resulting from the action of these forces, which, in addition to elevating them several thousand feet, have forced up through and between them vast masses of hypogene rocks, have obliterated every trace of the original lines of stratification.

It is a remarkable feature in the structure of slates that, however much they may have been disturbed during their period of upheaval, they assume the slaty structure in continuous or parallel lines extending over considerable distances. In the present case, they were found to extend without much alteration in their direction for more than seventy miles, between the Yuba and the Calaveras rivers. The angle of these plains with the horizon is very considerable; in fact, they are nearly vertical, having a slight inclination upwards towards the east. There is a coincidence between these and the slates of North Wales, in these respects. It has been observed over extensive districts in that country that the same parallelism prevails; and, in the direction of the laminæ, is near that of the magnetic meridian. Similar examples exist elsewhere.

The cause by which this is effected is somewhat obscure; but, from results of experiments made by subjecting masses of clay to electric currents, it is by some attributed to such currents on a grand scale beneath the surface. Professor Sedgwick suggested that the occurrence could " only be accounted for upon the supposition that crystalline on polar forces acted on the whole mass simultaneously and with adequate force."

This regularity, so far as was observed, did not seem to exist in all parts of the *coast range* where slate rocks were seen; which is probably owing to the fact of that region having been disturbed (by the causes which produce earthquakes) since the slates were raised up and assumed their fissile structure. But general conclusions in this connexion cannot be satisfactorily arrived at without a more minute investigation.

There is a tradition among the Indians to the effect, that the narrow strait which connects the ocean with the bay of San Francisco was effected within a recent period, (about four generations ago;) and that it was *suddenly* done. It is known that much greater effects than the sinking of a sufficient area to have opened this narrow passage have been produced by earthquakes within the last two hundred years.

The coast range bears incontestable marks of being within a district wherein internal disturbing forces have acted with violence sufficient to produce earthquakes of great energy during *very recent* geological periods; and there is no reason to suppose that these *apparently* slumbering causes may not at intervals manifest themselves, by producing visible changes upon the face of the country. Its geological structure indicates that it is within the great region of volcanoes and earthquakes, which embraces nearly all western America and a large area of the Pacific ocean and western Asia.

V. GOLD REGIONS OF THE SIERRA NEVADA.

This district of country, which seems of late very forcibly to have attracted the attention of the civilized world, certainly possesses importance, entitling it to a separate chapter.

Its auriferous developments have not only caused the application of the name " El Dorado," but seem to have called up the very spirit of romance which, three centuries since, conceived the idea of it, if we may judge from the many fanciful stories that have been promulgated within the past year.

Experience, however, has shown that it is safest to consider all subjects having relation to minerals (gold not excepted) in a purely physical point of view. It is *not* a proper field in which to give rein to the imagination. I shall, therefore, endeavor to treat the subject in hand as a case of the practical application of geology and other branches of science.

One writer, more than a year since, suggested the gold to have been ejected from volcanoes, and scattered far and wide over the country. This is so little deserving of attention, that it would not have been recalled, if, in the romantic excitement of the times, it had not been spread, and even believed, farther and wider than it purports to have scattered the metal itself.

It will, however, be dismissed with very few remarks, because it has not a single fact to support it; on the contrary, *all* the testimony is against it.

1. There are about three hundred volcanoes, including Solfataras, now known to be what is called " active," and a far greater number extinct, any of which may again become active, like Vesuvius (in A. D. '79,) which, after slumbering for a period reaching farther back than its history or traditions extend, again burst forth, and continues to burn.

2. It has *not yet been observed* that gold is among the products of volcanoes; although, if a volcano should burst through rocks containing gold veins, there might be metal mixed up with the fused matters in proportions so small, most likely, as to elude the search even of the chemist.

3. The wildest romantic spirits would not suggest as a hypothesis that *these* auriferous volcanoes vomited forth nothing but gold. They even would admit that a reasonable proportion of fragments of ordinary volcanic products accompanied the gold in its *flight*.

Now, between the Yuba and the American rivers, a distance of more than seventy miles on our line of travel, wherein so much gold has been obtained, there is a total *absence of all such volcanic fragments*.

4. The greatest horizontal distance that any fragment has been known to have been thrown from a volcano does not exceed nine miles.

5. Gold abounds among the detritus in the slate valleys near the Mokelemy river, it is true; but it has not been seen mixed with any of the volcanic products of that vicinity, unless it be such as may have been washed into the ravines with the adjacent stony matters.

6. Gold occurs in divers countries under precisely the *same* circumstances existing in California. On the Atlantic slope of the United States it was first discovered in the State of Georgia (where there are no volcanoes) among the transported stony matters in ravines, and subsequently traced to the vein in slate, in which it is known to exist as far north as Maryland. Larger pieces have been found in North Carolina than have yet been actually seen in California; and whether the comparatively smaller amount produced is owing to the destruction of a less amount of rock *and its included veins,* or whether the metalliferous veins are fewer in number and less rich than those in California, we have no means of determining.

In the preceding chapters the positions of the deposites containing gold, and the cause which produced these deposites, have been referred to whilst applying *established* geological theories thereto.

When I travelled through the gold region, no instance had occurred of gold having been seen in an undisturbed quartz vein; but geological inference so fully warranted the *certainty* of thus finding it, that at more than one locality I suggested a halt for a few days in order to clear the matter up by uncovering the outcrop of spots where I felt confident the metal bearing veins would be reached by a day or two's work; but the General's duties would admit of no delay, and we passed on.

Two months later, at San Francisco, Colonel Frémont informed me of his having actually removed the outcrop from a vein of quartz, which he described as being in slate-rock near the Maripoosa river, and showed a specimen of the quartz containing gold from the river—thus verifying the correctness of previous inferences. The specimen consists of a fragment of angular and *not water-worn* quartz, much discolored by peroxide of iron, which has undoubtedly resulted from the oxidation of sulphuret of iron that formerly filled the cavities left in the quartz. There was a large proportion of gold disseminated in small masses throughout the stone. This was taken from the vein near the outcrop; but it will be found, on penetrating beyond the influence of atmospheric action, that the gold will be usually accompanied by iron pyrites.

This is the only vein, so far as I learned, that has been actually opened; but the fragments of quartz throughout the gold region prove the veins to be *exceedingly numerous*, and in many instances of great thickness. Should the proportions of gold continue as great in this vein as it is believed to be by the owner and confirmed by the specimen, it cannot but prove highly productive if skilfully and econominally worked.

It is not a little amusing to read the romantic effusions of sundry letter-writers about this mine. A very recent one in the papers informs us that Preuss, and others of the party who went out with the Colonel, are working it; and that the gold is so plentiful that "they are already rolling in affluence, and wield an influence more potent than the wealthiest Atlantic nabobs." Now, the Colonel informed me that he did not expect to work the mine until government could secure the property from intrusion; and as to Preuss, (an old acquaintance,) Colonel F. stated he was exercising the duties of his profession in surveying land !

It is not to be expected that the quartz veins throughout the gold region will generally prove metalliferous: on the contrary, but a small proportion of the whole number can be expected to contain metals worth working, if we are to judge by the long experience in other metalliferous districts.

Although to a casual observer these veins may appear to have been formed without any apparent system or regularity, such will not prove to be the case whenever they shall be fully investigated. No surveys having been made of the veins in California, other countries must be referred to in illustration of them.

In all metalliferous districts hitherto sufficiently examined, there are several systems of veins traversing the same rocks, each system preserving the parallelism of its own veins with surprising regularity.

The means by which veins have been filled, is altogether an unsettled point in science; and it would be out of place to enter such a *wide field* of discussion as the theories hitherto proposed would lead to, but some remarks upon veins seem to be required.

Werner, the father of systematic geology, appropriately calls them "mineral repositories of a flat or tabular shape, which traverse strata (or masses) without regard to the stratification, having the appearance of rents formed in rocks, and afterwards filled up with mineral matter, differing more or less from the rocks themselves."

As an exception to this definition it may be remarked, that when such fissures are filled with trappean or other igneous rocks, apparently injected into them in a state of fusion, they are termed "dikes," and not veins.

Veins are usually nearly perpendicular; it is rare to find them inclining as much as 45°. Each system of veins in a district has its own direction or strike. Those first formed are intercepted by the next, and so on successively: their relative ages are determined by these intersections.

A knowledge of veins in mining districts is of so much importance, that they have been most attentively investigated, and nowhere more carefully than in Saxony and in England, where metallic veins have been pursued from time immemorial.

In the Freyberg district of Saxony, we are informed, there are eight distinct systems of veins. Of these, the first and oldest run chiefly north and south, and furnish mainly lead and silver. The second system runs in a direction nearly northeast and southwest, and are more argentiferous but thinner. The veins of the third system run north and south, and are crossed at right-angles by those of the fourth system, both containing galena. The remaining four are unimportant.

In Cornwall there are also eight systems, of which the two first contain little else than tin ore. Their direction is east and west, and nearly parallel, but the first are "cut" by the second at many points. The third run also nearly east and west, and cut the two first when they meet. In these are the great copper mines of the district. The fourth also contain copper, and run northwest and southeast. The fifth run nearly north and south, and contain neither tin nor copper, but a small proportion of lead. The sixth system contains copper, and the seventh is barren, as is also the eighth, which contains only clay not yet consolidated. Examples might be cited in great abundance, but these are deemed sufficient to show that each system of veins possesses a distinctive character. The number of different classes of veins in a district is variable, from two or three upwards.

If veins were filled, like dikes, by the injection of matter in a state of fusion from below, we should of course observe, in some instances at least, that a small excess had overflowed and was deposited upon the surface. There are innumerable examples of such overflowing from dikes, but none from veins. This insuperable objection to the theory of injection is fatal to it, and many others equally untenable have been suggested. The theory of electric agency has attracted much attention of late years, and in an increased degree since the experiments of Mr. Fox were made, a few years since, in England. This distinguished gentleman has actually succeeded in forming well-defined metalliferous veins by means of voltaic currents, operating under circumstances similar to those supposed to have occurred in Cornwall.

The rents and fissures in rocks have probably been produced partly by the same causes that give rise to earthquakes, and also by the contraction of masses of heated rocks in cooling. These rents or veins are continuous for great distances, passing through various strata or masses of the different kinds of rock in their course, and usually to depths greater than the miner has yet reached, although the earth has been mined to the depth of near 2,000 feet. In thickness they are from less than an inch to many yards, and a thick vein sometimes dwindles to a mere seam, and again enlarges to a great thickness. The largest known is one near Guanaxato, in Mexico, varying in thickness from 25 to 50 yards, and the next in Derbyshire, England, which averages about 23 yards. The former, according to Humboldt, was worked for silver to a depth of 1,080 feet, and for a distance of upwards of six miles in a horizontal direction.

The experience of miners shows not only that different classes of veins have different contents, but the contents of the same vein almost always vary in proportions when the vein passes through a different kind of rock.

In most cases veins are richer in metals amid rocks of sedimentary origin than in an igneous formation; adjoining them, but near the junction of the two kinds of rock, there is an increased proportion of metal or ores. In Mexico and Peru there are some rich metalliferous veins in porphyritic rocks, but these instances are rather to be considered as exceptions to a general rule. It may be readily conceived that these *innumerable* veins constitute so important a feature of California in an industrial point of view, that a most careful investigation by competent persons should precede the execution of any plan for *permanently* disposing of them by the United States. It is *impossible* that the boundaries of lots for leasing or selling can be judiciously located until the systems of veins be previously ascertained, and their several dips and strikes, or directions, accurately determined.

There may be many " systems" of veins in California and Oregon—for the latter country no doubt contains its metalliferous districts also. It is probable that different systems occur on different sides of the Sierra Nevada and Cascade mountains. Indeed the numerous large volcanic peaks favor this presumption.

There was no evidence seen, during my reconnoissance, of the existence of veins other than those mainly filled with quartz, and gold is not likely to be found extensively in any other. But it is by no means to be assumed that others do not exist: on the contrary, a judiciously planned and efficiently executed survey may be expected to develop other classes of veins containing silver, copper, and other metals; perhaps, also, the veins next the granite axis of the Sierra Nevada may contain ores of tin.

It is easy to explain why the debris of quartz veins should be the most abundant, and their outcrops readily seen. Quartz is one of the most indestructible substances in the mineral kingdom—being very slightly acted upon by atmospheric agents. Most rocks, indeed I may say all, are acted upon with more facility than ordinary or vitreous quartz, and consequently this mineral is found often projecting out in relief, or lying in masses over or below the outcrop of the vein, if it be on an inclined surface.

Calcareous spar, fluor spar, and other matters which form the greater portion of many veins, (especially where lead, silver, and copper occur,) are destroyed by solution and disintegration with comparative facility, especially the first named.

The destruction of such *veinstones* takes place often more rapidly than that of the rock itself, where detritus then conceals from view the outcrop of the vein.

The circumstances under which gold is found, taken in connexion with geological indications before noticed, *fully* warrant the belief that the whole extent of the Sierra Nevada has on its western flank metalliferous veins in a belt of country varying in width from thirty to forty miles, at least in the portion between the Yuba and Stanislaus. The mountains of the coast range contain, perhaps, several metalliferous regions, separated from each other.

The small branches from veins, called strings by miners, have been found, in many instances, to have no great depth below the surface; but the principal veins have never been observed to run out, and therefore may be considered in their industrial relations to extend indefinitely downwards. It has, however, been proved by mining experience, that in general, veins become less productive in metals after *great* depths are attained below the surface. It has been before remarked, that ravines have been cut into the flanks of the Sierra to great depths. Taking *only* 1,000 feet as a mean, it may be assumed that not more than one-twentieth of the solid contents of the gold region has been removed to that depth, of which little is left, except the gold remaining among the drift in the ravines. Many of the boulders now left were probably rolled down during the "drift period" from the hypogene rocks forming the higher portions of the mountain, and where there is no gold. This is the reason why the inferior portions of the drift are so much richer in metal than nearer the surface.

We know that by all this denudation, notwithstanding the vast proportion left in the remaining hills, there was a large amount of metal eliminated and afterwards buried in the drift, where he that chooses to dig is almost sure to get some *little*, at all events   But that is not the only effect produced in the formation of these numerous ravines. Without them the mineral treasures of this region might have remained concealed forever. Besides, each ravine is practically a road, upon which the sun of heaven shines, and much more convenient than the shaft or tunnel of the miner, which he would otherwise have been obliged to make by the expenditure of an enormous amount of labor, and which oftentimes he makes in vain. But here his chances of success may be more certain, provided the preliminary examinations have been judiciously and intelligently made.

In most parts of the world where metals are pursued in veins, a large expense is incurred in pumping out the water; and this in many portions of California would be enormous, (from the absence of *coal formations* and scarcity of wood throughout much of the country,) if pumping be resorted to. But here again nature has favored the miner by making provision for drainage, provided the mining be not carried below the level of the adjoining ravines, which, as we have seen, give access to the vein several thousand feet below their higher limits.

If the adits be commenced near the bottom of the ravines, and continued very slightly rising, the water of course will flow out; otherwise a small draining "level," as the miners term it, must be carried under the work. In order to enjoy the full effect of these advantages in draining, man must do his part.

The laws regulating the mining region should prevent any one from impeding the drainage of adjoining mines, and the government should, by law, retain the right to make any provisions that may seem proper in the premises. In the mining laws of Spain, and the Harz mountains especially, will be found many regulations applicable to California.

Although a large amount of gold has been collected in California within the past eighteen or twenty months, yet, considering the number of persons engaged in "digging" for it, the average amount to each is far less than is generally supposed. This conclusion is forced upon the mind irresistibly, when the results of the actual experience of a large number of the operators are taken into consideration.

The newspapers frequently relate instances of the return of individuals with considerable sums of gold. Many of these are much overrated, and the far greater number obtained it by other means than digging with their own hands—one portion by honest trading; but much of the hard-earned treasure in the hands of returned individuals has been borne off in triumph, and brought home as the spoils of the conqueror, in contests where honor belongs to neither winner nor loser.

Representations from and about California are to be received with many "grains of allowance." The preternatural excitement which has been produced by divers causes, in some cases to promote individual benefit, has really impaired to a large extent the faculty of seeing things in the light they would otherwise have been viewed. And there is yet no prospect of an end to this state of things, because, as soon as the public mind begins to recover from the effects of previous causes of undue excitement, additional ones are presented in the shape of most exaggerated accounts of golden discoveries. Whether the public good will be promoted by this state of things may well be doubted. A reference to *some* of these causes it is proper to give.

It is the interest of the numerous traders within the gold region to collect around them as many "diggers" as possible, and each is very naturally induced to regard favorably "the diggings" of his own vicinity, and takes means to spread accounts of its richness. Wonderful stories are circulated in some instances to increase the population at a particular spot; and when the diggers flock to it they often find it no better than the one they left, and sometimes less productive. A very large proportion of those persons we saw in the gold region were *in transitu;* and upon inquiry we learned from them usually that the place they had left was unproductive, and they were bound for another, which they had *heard* was producing very largely; and on the same day, perhaps, would be seen other parties "*prospecting*," as they term it, or looking for better diggings than the poor ones they had left, and in many cases just from the reported "*good diggings*" the first party were going to. At some of these places you would hear of some one being very fortunate, and that they averaged per day a half, one, two, or three ounces; but, like the tariff for postage, they never appear to get $1\frac{1}{2}$, $2\frac{1}{2}$, $3\frac{1}{2}$, and so on. These accounts from particular spots sometimes find their way into California papers, and from them are copied and spread far and wide at home.

Notwithstanding all this waste of time, and that nine out of ten who left their homes under erroneous expectations in reference to the facility with which the gold could be had, have been cruelly disappointed, yet, the extent and number of the ravines containing gold is such, that the

Ex —3

large number of "diggers" have in the aggregate produced a considerable amount of this metal.

It is impossible to ascertain the amount of labor there hás been required, or, in other words, the average number who have worked at the diggings, and the number of days work of each. The maximum number has been estimated at 80,000 by some of the writers, which is, perhaps, at least 30,000 too many. If, however, we suppose only 10,000 to have worked steadily during three hundred days out of about six hundred since the digging began, and suppose each to have gathered an average amount of $3 per day, the aggregate would amount to $9,000,000, being very much more than the whole amount exported in *every way* from California up to the 1st December last, to all countries, Oregon inclusive. As the cost of living fully equals $3 per day, it would appear that gold digging is not as good as laboring at home, where the laborer can save something.

The mode by which the gold has been obtained hitherto, has been often described, and, except in a *very* few cases, is effected in a most unartistical manner. In general they work singly, but in some cases two or more work in company. In the drift on the running streams they have to work among the boulders and gravel, and are much troubled with water, which they have to bail out. A man digs among these materials usually without getting much of the gold until near or at the bottom of the alluvium. Often immense boulders interpose, which he is obliged to work around, and scratch out all he can from underneath. The sand with the gravel (from which the larger pieces of stone have been picked out and thrown away) is either washed in a tin pan with water, (to which a whirling motion is given, so as to throw out the matters other than gold, and the grains of specular and magnetic oxides of iron,) or in a cradle or rocker. The tin pan people usually work singly, whereas those with cradles work two or more together. There is unquestionably a large proportion of the finer particles of the gold lost by nearly all the operators I saw at work.

Most of the fragments of gold from the *large* ravines that I have seen bear testimony to having been scoured by the passage of stones over them, so as to have suffered a great loss in weight. The matter thus rubbed off was in a state of minute division, most of which lodged in crevices near by. Some, perhaps, so minute as to adhere to the passing sand, &c., has been deposited lower down, and in widened places, where the lessened velocity permitted the deposite of sand bars.

There is no practicable and economical method of retaining this finely-divided gold known, other than by the use of mercury, which forms an amalgam with it. The amalgam, having about six times the specific gravity of the gravel and sand, remains in the rocker, whose inclined position, and brisk alternating motion, with the aid of the current of water, get rid of the earthy matters—the gravel being separated by means of sieves. In working without mercury, the gold is always obtained with the grains of iron ore, the complete separation of which is usually effected by blowing it out with the breath, but always with the loss of an *additional* proportion of gold. Mercury does not act upon these ferruginous grains, and consequently the amalgam contains none of them; but mercury will take up platinum, silver, lead, &c. The amalgam is put into an iron retort, with a closely-fitted beak, and exposed to a heat sufficient to drive off the mercury, which is distilled over, and condensed in a receiver for a subsequent use. To be sure, however, that it is all driven off, the remaining gold in the retort should be fused in a crucible.

A most efficient rocker was in use in July last, at Mormon island, of the kind so much approved at the Virginia gold mines. It was constructed so as to prevent the waste of mercury or gold whilst in it, and doubtless will enable the operators to separate all the gold from the other matters. This machine is so superior, that it was rapidly coming into use, and may be considered the first step towards improvement in California in the art of separating the gold from foreign matters which it is mixed with as it is procured from the deposites of alluvium.

The bad success that attends the efforts of much the larger portion of the operatives must soon materially lessen the number of those working independently, or in small parties, because of the inherent difficulties in this mode of operating, which is nothing more than digging holes here and there, and throwing the rubbish about, so as to interrupt the digger himself, or his adjoining neighbor, in subsequent operations, very materially.

A year or two more will suffice to exhaust most of the metal which is readily accessible; after which, a prize will so seldom be met with, to sustain the hopes of the poorly-rewarded gold-digger, that he will find it his interest to work at moderate wages for those who are possessed of the requisite means, skill, and knowledge to manage the business *"secundum artem,"* and provide comfortable homes for those whom they employ. This, of course, cannot be done until proper legislation shall enable them to secure the property in such manner, and with a sufficient amount of land, to operate under the most efficient and economical system.

When that shall happen, most of the ground which had been previously scratched over will be *systematically* worked again. The large boulders (so much in the way) will be removed by blasting, or hoisting, or other means, according to circumstances, and little, if any, gold will escape; and it is more than likely a greater amount will be obtained by the second washing than had been previously.

In the mean time, an investigation of the veins will be in progress, so that it may be ascertained whether some of them may not prove worth working. That they are exceedingly numerous, and many of ample size, has already been stated.

The experience of other parts of the world hitherto is very unfavorable to the prospect of profitably working them. In Mexico, they are rarely worked. In Peru, where such immense quantities of gold were obtained by washing, in days long past, the veins have proved unprofitable in almost every attempt to work them. The same may be said of Chili. In the Uralian mountains, the Russians get gold to a large amount from washings done under government direction, and in the most approved manner. There has little or nothing been done in mining gold in the veins of that region, and the published accounts speak only of metal from the washings.

The gold region of the Sierra Nevada embraces about three-fourths of its western flank; and within this large area, it is probable that few of its countless ravines do not contain some gold. But it is only in a limited number of these dry and smaller ravines in which gold is known to exist in sufficient abundance to entitle them to the California cognomen of *"dry diggings."*

The expense of transporting the matters from which the gold is to be washed to water, or the water to them, prevents, at present, any but the most productive of them from being worked. All the metal in these ravines has been derived from the destruction of those portions of veins that once extended through the rocks which occupied, between the hills, the *spaces* now constituting the ravines.

It would seem, therefore, correct to infer that the quantity of gold among the detritus of these ravines must bear a relation to the richness and number of the veins containing this metal, which will be found, by a careful survey, cropping out within the limits of the ravine, generally concealed by detritus.

Large amounts of gold have been procured in certain parts of but few of these ravines, which may prove either the existence of one or more veins containing a large proportion of metal, or that there may be connected with the ravine *many* veins, none of which will pay the *enormous cost* of eliminating the gold therefrom—a very common occurrence elsewhere, notwithstanding the *brilliant* prospects that have been sometimes presented at the beginning.

There are thin pieces of gold often met with, which many suppose, erroneously, to have been "flattened out by pressure." It is not uncommon to find in metallic veins films of metals in their native state filling thin fissures in the veinstone. Most cabinets contain such pieces (often in the matrix itself) of gold, silver, and copper. Those of gold from California have in many instances been smoothed by *attrition*, after being separated from their former position in minute fissures of the quartz.

Should veins containing gold be developed and fully *proven* to be valuable, they can only be profitably worked by a proper combination of skill, capital, and machinery, under good management. The underground operations *must* be carried on in a regular mining fashion; and, after separating the refuse matters called "deads" from those containing metal, (which requires acute judgment,) they must be ground sufficiently fine to enable mercury to dissolve out all the gold. The varied operations requisite make in the aggregate such an enormous amount of cost, that there are few gold-producing veins in the world containing a sufficient proportion of metal to meet these expenses.

VI. THE QUICKSILVER MINES.

Although the "quicksilver district" (as the vicinity of the mine whence that metal is procured in California is termed) was not examined, yet the applications that must necessarily be made of mercury in California is a subject so important as to render some reference to it very proper.

From what has been already said, it may be inferred that, whether gold or silver *veins* be extensively mined or not, there will be required a very large amount of mercury in California for separating the gold obtained in the ravines. In view of this, it would appear the duty of the government to legislate in reference to its mines of cinnabar so as to insure the *greatest possible* product of a material *so essential* as an industrial element in California.

The most important quicksilver mines in the world hitherto have had, and continue to have, their products monopolized by one great trading concern, which regulates the prices at will. We should jealously guard those belonging to the nation from a similar fate.

It is believed, if the value of mercury could be reduced to what it was previous to the great advance in price some years since, the product of precious metals in Mexico alone would be nearly or quite doubled. The richest mines only can be worked in that country under existing circumstances.

Under Spanish dominion, the regular supplies of mercury from the mines was considered so important to the trade of the kingdom that the government took it in hand. They furnished and delivered quicksilver to all parts of Mexico at one *fixed price*, so that those conducting mines might make their calculations accordingly. By this means operators were certain of supplies of this essential material at fixed rates; and thus was induced the working of a great number of veins whose rates of productiveness and distance among the almost inaccessible mountains were such as would have otherwise prevented anything being done with them.

It is not to be supposed that our government will take this method of furnishing quicksilver. It would, however, appear incumbent upon it to adopt efficient means, without unnecessary delay, to cause a most minute and careful survey to be made of the region within which cinnabar may be expected, which is situated between the " great" or Sacramento valley and the sea—being, in fact, the area occupied by the many ridges of the coast range, with their numerous intervening valleys.

The mines of cinnabar at *present* known, or which may be discovered in the progress of the mineral survey suggested, being the property of the United States, should by *no means* be permitted to pass from under the full control of the government until a *well-matured system* be devised, which will not only insure a production of this important material *equal to the full extent of the capacity of each mine*, but which will also secure them, or their products, against the possibility of being monopolized by any one private concern, or by any combination that may be formed for such purposes.

The very few productive quicksilver mines in the world, and the fact of the actual necessity for its use in reference to the other precious metals, present strong temptations to monopolize them, which has been proved a very practicable operation.

In October last the price of mercury rose to $5 per lb. in San Francisco, and even at that exorbitant rate it was advantageously used.

VII. OTHER MINERAL RESOURCES AND THEIR INDUSTRIAL APPLICATIONS.

There has been so little done towards investigating the mineral resources in general of California and Oregon, that our information in reference to them is exceedingly meagre.

The interest generally felt in reference to the subject of gold has occasioned it to have been dwelt upon at some length in the preceding pages, with the view of endeavoring to give some idea of the actual condition and future prospects of the branches of industry its discovery has created in Upper California.

Mercury has also been considered; but chiefly in reference to the important relation it bears to the production of gold, silver, or platinum.

The silver mine near the southern terminus of San Francisco bay is said not to be rich; but upon investigation this may prove otherwise. Aside from the occurrence of this mine, geological indications, taken in connexion with those of other silver-producing regions, give grounds for believing that veins containing silver or its ores may be expected both on the coast range and on the flanks of the Sierra Nevada. It may yet happen that California, like Mexico and Peru, will be more celebrated for its-silver than for gold, by producing it to a greater amount in value.

Copper ore may also be expected, as in Peru and Chili, to be found in ample supplies. The reason why the veins containing the two metals last named are more likely to have their outcrops concealed, has been explained—so that it must not be supposed they are absent because there is yet so little evidence of their existence.

Veins of lead ore are known to show themselves near Sonoma; others will doubtless be found in the coast range and in the Sierra, to swell the amount of products of these *metalliferous regions.*

There is, however, one drawback to the country, and a most serious one, in the absence of that main-spring to industrial operations in the present age—*coal* I had thought it not impossible that, if the Sacramento valley itself should not be ascertained to repose upon a " coal formation," one or more coal basins might exist among the coast range of mountains; but was disappointed in this *leading object* I had in view in visiting the country, by finding the strata not older than the comparatively recent periods of the tertiary formations resting immediately upon the hypogene rocks, thus showing the remarkable fact of the *total absence of the entire suite of sedimentary formations (from the tertiary down to the silurian)* which form the surface of the greater portion of the known world; and, as the *coal formation* has its place in the midst of these, it is of course also *wanting.* It would be premature to assert positively that it may not exist north or south of the regions covered by my reconnoissance; but, from all the information I have been able to collect, it seems likely that the same geological features extend from near the Oregon boundary to the southern terminus of Lower California.

Not having visited Oregon, I am not in possession of sufficient information to found an opinion upon the probability of the carboniferous formation existing in that Territory. Lignites and tertiary coals are known to exist, and it is most likely that upon each discovery of these a " new edition" of a " coal report" is emitted.

Reports of the discovery of coal beds, and very many other matters in California, have been told, written, and published largely.

Whilst Colonel Mason was in command there, he considered the subject so important that " he directed the late Captain Warner to visit and examine every locality in which coal was reported to exist" up to that period, now more than a year and a half ago. The official despatch of the Colonel gives the result; and I have also conversed with the lamented Warner himself upon the subject.

It appears that *every one of these numerous " beds of coal of the best quality for steaming"* proved to be either *lignite* or *bitumen*, or something or other still further removed from the character of coal. The lignites, in most cases, were but fragments of trees or single trees only.

There is an ample coal formation on Vancouver's island, and others on the continent further north. It is to *that quarter* that California must look, unless Oregon may produce it.

We are often told about coal on the west coast of America; and one who has not examined the subject would be apt to suppose that coal beds were "plenty as blackberries" from Bhering's straits to those of Magellan; but so far as I can learn they are all lignite or tertiary coal, and of little use, except for burning lime and like purposes, even where they occur to some extent.   The British Pacific Mail Company have caused examinations to be made at various places, from Costa Rica to Cape Horn, of reported coal beds.   Their agent thought he could secure a supply at one spot only, (Talcuahano,) and did some mining five years ago; but the first bed ran out after producing 5,000 tons of inferior *tertiary coal.*   So far as I have learned they use very little of this inferior material, and rely upon English coal almost entirely.

In consequence of the scarcity of fuel and its high price, the copper ores of Chili and Peru are generally shipped to Great Britain and the United States for smelting.

If this ore should be a product of California, it must, of necessity, take the same course; and thus give return freights, so much needed in the trade between that region and the Atlantic States and England.   The copper ores *must* go to the fuel *always,* because of the less amount to be transported than when the fuel is taken to the ore.

The probability of the existence of ores of tin in the upper portions of the Sierra Nevada has been already hinted at, and is based upon the fact of its occurrence in various other parts of the world, (especially in Mexico and Chili,) with mineral accompaniments similar to those of the Sierra.

Sulphuret of lead has been noticed near Sonoma, and I believe on good authority.

It is more than probable that available veins of this ore will be met with in the coast range, and the same may be said of zinc.

Indications of those valuable ores of iron called "specular, and the *magnetic* oxides," were seen, and reference made to them, whilst giving sketches of the minerals on the routes of travel.

As there are magnificent forests covering the flanks of the Sierra Nevada, from the summit, for many miles below, it is possible, in *after years,* there may be found large veins or beds of these ores, (as in Sweden, of magnetic, and in Elba, of specular oxide,) and that the manufacture of iron may be advantageously prosecuted in this *elevated region,* where there is ample water-power and fuel.

From the higher portions of this timbered district it is unlikely the wood or timber can be transported to supply the wants of the valley, but will remain scarcely touched until it can be profitably used within the district itself.

A remarkable feature in the geological structure of California is the *innumerable* localities of serpentine, with its associated steatite, or soapstone, and other magnesian minerals.   Serpentine here, as everywhere else that it occurs, is an intrusive rock, and differs from most others of like origin, by occupying at the surface *isolated* areas of an irregular rounded shape, with diameters from one hundred yards to several miles, and rarely, if ever, in the continued lines which trappean and other fused rocks are often observed to affect.

It is believed that all the magnesia and its sulphate (or epsom salts) is now made from the silicates of magnesia obtained from the formations of serpentine.

The valuable preparations called chrome yellow, (chromate of lead,) so extensively used as a pigment, and the bichromate of potash, which forms an important material for the dyer and calico printer, are manufactured from an ore composed of oxide of chrome and oxide of iron, which hitherto seems to have been obtained exclusively from beds or nests in serpentine rocks.

The extraordinary abundance of these rocks in California, compared with the few isolated spots which they occupy in Europe and eastern America, justifies the expectation that they will give rise hereafter to an extensive manufacture of the articles above referred to. The probability of this is greatly increased from the fact, that most of the additional materials necessary to these branches of manufacture will be obtained in California at much less cost than in the Atlantic States or in Europe. These are the nitrate of soda from Peru, nitrate of potash from India, (and perhaps from California itself,) sulphur from the numerous volcanic islands of the Pacific, and quite probably from some of the extinct vocanoes of Upper or Lower California. Potash may be produced from the forests of the Sierra Nevada, as well as from those which clothe most of the coast range, from the Russian river, near Bodega bay, to Puget sound. The remaining material, lead, has already been treated of.

One cannot but be surprised at the apparent scarcity of *limestone* within the portion of California embraced within the range of my reconnoissance.

At one place only it was met with, upon the flanks of the Sierra Nevada, not far from the Cosumes river, where there exists in small quantity an impure variety of crystalline dolomite, or magnesian limestone. But it is by no means to be inferred that a survey would not prove the existence of limestone in numerous places, because it is very common to find it interstratified with metamorphic rocks in beds of no great thickness, whose outcrops are often concealed by causes similar to those referred to when treating of veins containing calcareous matters.

In the coast range, at Livermore's, a rich marl was met with, which produces lime, upon being calcined; and the large fossil oyster shells, being abundant, are used for the same purpose.

The thin beds of limestone on Suisun bay, near Benicia, and their application, complete the *resumé* of calcareous localities; and, although it may be stated that no extensive developments of limestone are likely to exist in this part of the coast range near the bay of San Francisco, or those connected with it, yet it would be very remarkable if these sedimentary formations, so eminently calcareous in many parts of the globe, were not so, at least to a limited extent, in some portions of this range. Besides, the metamorphic rocks of the Coast mountains may be expected to contain crystalline limestone.

But, if it should turn out that the land will not yield a supply of lime, the sea will make up the deficiency.

Shell-fish of certain kinds are exceedingly numerous upon these rocky shores of the Pacific, and around the numerous islands contiguous thereto. Among these is the *haliotis iris*, (often nearly one foot long) — an article of export for some years past for ornamental manufactures; but of those unsaleable for such purposes, and of other shells, an ample supply exists along the coast.

Taking into consideration the climate and other circumstances in connexion with California, it seems probable that lime is not likely to be used very extensively as a building material. The exorbitant price of lumber and many other matters will not continue long. The frightful destruction of life and health, the severe trials, exposures, and discomforts, attending the *present* mode of getting gold, coupled with the small return which much the larger number receive, will drive them to other pursuits, and, among these, to furnishing lumber from Oregon, as well as from the extraordinary forests in California, to be hereafter spoken of. In a very few years, lumber will be nearly as cheap upon the tide-waters of California as in the Atlantic States, so that it will be largely used for building purposes.

There is, however, a still cheaper material for house building, in connexion with lumber. From the first settlement of California, the *adobe* was and is yet used. This *sun-dried* brick, the knowledge of which the Saracens carried to Spain, the Spaniards caused to be made, as the Egyptians did some thousands of years since: the only difference is in the fact that the Jesuit fathers in California permitted the Indians to use grass and straw, whereas the sons of Jacob were required to use the clay *per se*. These adobes are usually made from eighteen inches to two feet long, twelve inches wide, and five or six inches thick, but contract considerably in drying, unless chopped straw or hay be previously mixed up with the tempered clay. This large brick is easily broken.

The walls are constructed of these by the Indians; and their surfaces are very smoothly plastered over with clay, with the aid of a rude wooden substitute for the plasterer's trowel.

The most comfortable houses I saw in California were thus constructed. Some thought that, like stone and brick houses, they were damp; but no *evidence* of this appeared in the condensation of moisture: on the contrary, the internal surface of these walls was always dry. The thickness of them, in connexion with the slow conduction of heat through dry clay, keeps the temperature in the interior very uniform. Consequently, when the temperature of the external air rises through the day—as it frequently does in some parts—thirty to forty degrees, it may contain more moisture than can exist in it, when cooled within the house: so that, perhaps, the air feels *chilly*. A very little fire made up once or twice a day, the year round, would remedy this inconvenience, where it occurs; and this will be considered essential for the comfort and preservation of health within ten or fifteen miles of the coast, (north of Monterey,) whenever the social state of California shall progress to the point of esteeming individual comfort and health worthy of attention.

It would be strange indeed if our countrymen could not contrive a better means of constructing houses of adobes than was practised by Moses and his hard-tasked kindred; and this, in my opinion, they have done—namely, by tasking the steam-engine.

The application of labor-saving machinery to the manufacture of bricks has occupied the attention of many intelligent minds in Europe and America for a long period, and some excellent results have been produced various inventions, some of which are applicable to particular stages of progress, whilst others are intended to receive the clay as it is dug, and produce therefrom the moulded bricks, without other manual labor than consists in supplying the clay and taking away the bricks.

Bricks moulded by any of those machines that will give them *sufficient tenacity* and regularity in form, or such as are well made by hand, may, without being burnt, safely be used in California, west of the Sierra Nevada, and south of the 40th parallel of latitude, because of the absence of any frost that can possibly injure them.

With the aid of machinery, they may be moulded and dried in the sun, at the present prices of labor, for one-tenth the sum that kiln-burnt bricks now sell for in San Francisco. The material for this manufacture, from the most loamy to the most tenacious, abounds throughout the country westward of the Sierra Nevada, especially in all the valleys of the coast range, and in the Sacramento valley.

No lime is needed in using these sun-burnt bricks. The mortar made use of consists of clay mixed up with water, of which a very thin stratum would be needed with well-made bricks; but the rough adobes of the country require a large proportion of this clay mortar to make a tight wall. The foundations should be carried, as in framed buildings, a foot or so above the ground with burnt bricks or stone; and, the walls being smoothed off with water, a cheap, durable building for the climate results, calculated to preserve a uniform temperature within, and promote the comfort of the legitimate occupants, whilst it materially abridges the accommodations of certain little leaping and biting crustaceæ which are so annoying about most domiciles in California, and where they very much exceed in size, activity, and voracity their brethren elsewhere.

There is not sufficient frost to produce the slightest effect upon these walls; and the winter rains do not injure them in the least, if properly smoothed and whitewashed, provided the roof sufficiently covers them. As an additional precaution, instead of whitewashing, their external surface might be coated with a mixture of sifted clay or earth and the bitumen, so abundant in California: these should be melted together, and applied after the walls become dry. The internal surface may or may not be similarly coated, unless papering or painting be preferred.

In view of the importance of this subject, in its relation to public as well as private construction, it has been entered into somewhat in detail. It is of especial importance to the country in general, but more particularly in the Sacramento valley, where there is no stone and little timber, as well as in the settlements and towns within the coast range—such as San Francisco, Benicia, &c.

The sandstones and other rocks of the coast range will furnish ample supplies of building-stone, much of which is firm and strong, like that noticed near Benicia. In some places it is soft, and would soon crumble to sand in a wintry climate, but would be more durable here, from the almost entire absence of the principal causes of destruction elsewhere. These are: the freezing of absorbed water, which, by its expansion, separates portions of the stone; and the growth of moss, which extends in damp weather between the grains of porous stones, and lessens their cohesion, by abstracting, for the growth of these plants, the lime, alkali, &c., contained in the cement. The carbonic acid also, when water is present, lends its aid in dissolving out the same matters.

The examination of the numerous *mineral* springs in California, although a subject of great interest, could not be attended to during my journey through the country. The properties of these springs should be investigated by analyses and otherwise, especially those from which the water issues at elevated temperatures.

It seems to have been assumed by many persons that all the more valuable kinds of gems must necessarily be found in this country; and there are those who were probably so sure of it, that they thought it as well to announce the finding, by way of anticipation, "before the fact."

The geological character of the country would certainly permit the existence of all the varieties of what are termed "precious stones;" but there was no reliable testimony of any being found, except diaphanous crystals of quartz, (which are often mistaken for diamonds,) and the compact varieties of this mineral, such as jasper, chalcedony, and agate, and the hydrated variety, among which is opal. Of this last, the much sought after description, called "fire opal," is found upon the southern part of the western slope of the Sierra Nevada. It differs in composition, structure, color, and appearance most widely from emerald; and yet it is the identical stone about which so much flourish has been made in the papers, under the name of that costly gem, the emerald.

VIII. VEGETABLE PRODUCTIONS, (INDIGENOUS.)

A very large proportion of this country is wholly destitute of trees of any kind, but there are very extensive districts containing dense forests of trees of enormous size.

The hills with which the flanks of the Sierra Nevada terminate are too arid to support trees, and but little of anything else (except during the rainy season) until after a considerable elevation is obtained. The valleys, through which the *constant* running streams pass, upon approaching the edges of the Sacramento valley, are dotted, in many places, with open groves of oaks, which, wherever the soil is more than usually rich and sufficiently moist, attain a large size; in other places they remind one of well-grown apple trees.

These oaks lessen in number and size above the points where these valleys change to the narrow ravines through which the streams flow from near the summit of the Sierra Nevada; and finally give place to other kinds of oaks, one of which appears to be identical with the black oak of the Middle States; others approach in character the red oak.

At an elevation of about 1,000 feet, pines begin to appear; the first seen was the kind bearing cones about the shape and size of the largest pineapples. We observed no large trees of this species, and at 50 or 60 feet above the ground they terminate in open straggling tops. Unless they attain larger dimensions further up the mountains, they can be of little importance for timber.

The large tall pine (P. Lambertianii,) with seed cones 12 to 14 inches long, begins to appear at a higher level, and, as with the others mentioned, the size increases at greater altitudes.

There are several varieties of cypress and cedar, forming stately trees, the more elevated country; and it is quite probable there are also others, useful in the arts, which escaped notice.

I have been informed, that higher up than we reached, (especially in the moist region below the snow line,) there are other varieties of trees. appears that not only the size but the number of trees increases with the altitude of the slope of the Sierra. The largest of the pines was not less than 220 feet high, and upwards of five feet thick a few feet above the ground, and, like most of its species, ran up, to all appearances, fec⸱ly straight.

There is a very large amount of timber in this region sufficiently sible to furnish supplies, for agricultural and other purposes, to the lo country, whenever proper attention shall be turned to making roads.

The oaks which were noticed in the small valleys also usually line th water-courses of the Sacramento valley, at least such as are not border by marshes. In many cases there are strips of land covered with oaks the banks of the large rivers within the valley, separating the mars from the rivers.

These trees furnish a hard tough wood, whose properties are very si ilar to the white oak, and is well adapted to the manufacture of agric tural implements and machinery.

The climate of the southern country does not seem to have permit the growth of forests except upon the high ridges, where are stated to extensive forests of pine and the palocolorado or redwood. This appears to be larger further north, where it grows at a lower level, but tains its greatest size in about latitude 37°. Northward of the bay San Francisco it begins to approach the tide level, and, after passing latitude of Bodega bay, its forests, intermingled with pines, oaks, other valuable timber trees, cover the larger portion of the country tween the Sacramento valley and the ocean. The lessened size number of this valuable tree, as it approaches the Oregon boundary, compensated for by the reverse taking place in regard to the pines wh seem to replace them to the northward, where they attain the enorm dimensions given in the narrative of the " exploring expedition."

The extraordinary dimensions of the redwood, its great abundan and its value as timber, entitle it to an especial notice. It is a cypr with a lofty trunk, from which, often, there is not a single limb or bran within upwards of one hundred feet of the ground. The grain of t wood is so straight, that, in the absence of saw-mills, most of the plan formerly used were obtained by splitting them from the trunks of this tr The wood is of a rather dull red color, with a specific gravity and stren about equal to that of the white pine, but it is stiffer.

It is very durable, as is evinced by posts seen by Colonel Frémon " which had been in the ground there three-fourths of a century, with traces of decay." The houses built by the Russians, many years sin at Bodega, are of this timber, with posts sunk into the ground, into wh the horizontal pieces are mortised. The planks upon the sides are ab three inches thick, and had been but little smoothed off after being s No signs of decay could be detected even in the posts. At this p Captain Stephen Smith erected a steam, grist, and saw-mill, (about years since,) being the first structure of the kind in the country. states they cannot saw logs much over four feet in diameter, and nev cut any of less than eighteen inches; and yet the average number of c or logs, each of sixteen feet in length from a single tree, lies betwe eight and ten.

Westward of the bay of Monterey, where the redwood attains its grea est size, Frémont describes the largest he measured to be fifteen feet in diameter, and two hundred and seventy-five feet high. Lieut. Stone man, United States army, who has been some time in the country, re lates that another, of about the same height, was twenty-one feet in diameter.

The largest tree which I saw in the forests near Bodega had been cut down by the Russians, and much of it removed. The stump was twelve and a quarter feet in diameter, clear of the *very thick bark.* There were many trees nine or ten feet, and those of six to eight feet in diameter were very common.

After examining these forests to some extent, I measured off a space equal to one seventh of an acre, which was estimated to contain about an average of the forests of that region, and found within it three trees about one hundred feet high and eighteen inches thick, and twelve others varying between four and eight feet in diameter, and from 180 to 230 feet high.

It is difficult to realize an idea of the product of timber upon an acre containing the proportion within the fractional part above noted without an arithmetical calculation, when it will be found to produce about one million of feet of boards one inch thick, besides five hundred cords of wood from the tops and limbs.

Captain Smith thinks he alone has ten thousand acres of such forest, and I saw many acres which would yield considerably more in proportion than the measured space.

One is not more surprised at the size of these trees than at their being so crowded together. They grow usually in very close circles, which seems to be owing to the sprouting of shoots from trees that have decayed or been blown down.

It was observed in the forests near Bodega, where timber has been cut for a number of years, (formerly by the Russians,) that young trees always sprout up around the base of the stumps, as is the case with chestnut and other trees of the United States.

The redwood tree possesses so great a value for fencing as well as building, that efforts should be made to propagate it in many other parts of California where it may be grown. I regretted that it was not the season for seeds, because there is little doubt it would thrive in many of our States ; and it is to be hoped that ample supplies of seed may be introduced by public or private means.

Oaks, of the several kinds which were noticed in the Sacramento valley, and upon its eastern borders, are sparingly distributed upon the portions of the coast range and its valleys that I traversed. A wagon maker near Bodega stated that its strength and durability equal that of our best white-oak, which it resembles more in internal structure than in fruit, leaves, or trunk.

Some of the oaks produce a superior bark for tanning leather, and, what is remarkable, do not *appear* to suffer from being *entirely* stripped of their bark from where they branch off (at six to twelve feet high) to the ground. At the end of four or five years the same trees again yield a full supply of that portion of the bark in which the *tannin* resides, when they are again stripped.

Trees were pointed out to me whose trunks had been completely deprived of their bark at three several times, commencing with the occupation of the Bodega region by the Russian Fur Company.

There are several other kinds of trees in the northern parts of the coast range which deserve examination: among them is one having a *very hard close-grained* wood, which appears to be well adapted to cabinet-work, and for certain purposes in machinery. Its smooth bark comes off in thin sheets, like some varieties of the birch, and exposes a fresh layer of bright yellow color, which passes through the various shades of yellow and red to brown in winter. In the Sierra Nevada it is a large branching shrub, but on the coast range, near Sonoma and Bodega, it becomes a tree whose trunk is three feet in diameter. It is used for making the wooden stirrups of the country, and other purposes requiring a hard strong wood, not apt to split. It is capable of being polished so as to present an exceedingly smooth surface. There are other trees; but as it is only intended to notice those of most importance for industrial purposes, they will be passed over.

Other vegetable products, not embraced within the range of my investigations, will be referred to those who are skilful in botanical researches. I brought home with me specimens of the bulbous root called the amole or "soap plant," and shall take an early interval of leisure to make an analysis of some of them, in order to ascertain to what its detergent properties are owing. It is certainly as effective as any of our artificial compounds used for similar purposes.

In the Coast Range mountains there are found currants, gooseberries and raspberries, the last two similar to the indigenous varieties of the Atlantic States. The currants are of the black variety. Strawberries are also indigenous to this region.

The coasts of California and Oregon abound with marine plants, the most important of which is the one ordinarily called kelp, (a species of fucus,) highly productive in soda, which was formerly obtained exclusively from the barilla or kelp, produced by burning the dried fuci. Of late years, however, it is obtained more cheaply from chloride of sodium or common salt. These plants yet constitute the cheapest material for the manufacture of iodine, so much used in medicine. As a manure they are also greatly prized, especially on dry soils like those in California.

The lichens, so plentifully covering the rocks near the tide-level along the coast, may prove worthy of attention, like the two varieties of the plant from Iceland and Ireland, improperly called mosses.

IX. ANIMALS INDIGENOUS TO CALIFORNIA.

The mineral and vegetable resources of California having been considered to some extent, a brief reference may now be made to the animal kingdom.

The most prominent four-footed native of the country is the bear, of which I had not an opportunity to see a living specimen; but as this animal has had a full share of attention from several who have written from or about California, he need only be named.

There is an animal often spoken of in the country, and which will attack most ferociously horses, cows, and all other creatures except man and the bear. The name of "California lion" is applied to this animal very improperly. From the description related, it appears to be an approximation to the panther. They are not at all numerous.

The common gray wolf inhabits this country, but not in great numbers.

The cuyote abound throughout "hill and dale," and are very destructive in killing sheep and pigs. They will be rapidly destroyed upon the introduction of dogs, without affording much sport to the fox-hunter, because they are by no means a fast-running animal. In size they are between the fox and the wolf, and resemble both.

Foxes are sometimes seen, and among others the silver-gray variety, so much prized for its fur.

Large herds of elk are often seen, and the abundance of their antlers strewed over many of the plains shows that they must be, or have been, very numerous.

Antelopes do not, probably, exist in as great numbers as elk, but they are frequently seen in small herds.

The wooded regions abound in deer, especially northward of Bodega, where, in the little glades among the redwood forests, they may be met with at any time.

Rabbits and hares, the latter very large, and with ears of great length, are often started up in passing through many parts of the country.

The only squirrel met with is one about half the size of the gray squirrel, and not unlike it in appearance. They burrow in the ground and abound in the groves of oak trees; but in one district, in which there were immense numbers, (in the coast range east of San Francisco bay,) we did not see a single tree for thirteen miles; and as the surface of these parched hills did not seem to afford visible sustenance for such vast numbers, it is probable there may be certain *nutritious* roots of the dried plants upon which these squirrels live. It might prove both interesting and useful to determine the point. The district is traversed by the road from Stockton to the San Juan valley, and embraces the first 13 miles of the coast range on the coast. The whole sides of the ravines through which we passed seemed to be undermined by these animals.

Land birds are not found in great variety in California, and the traveller misses the delightful notes of our singing birds. Some kinds, however, are very abundant, especially the partridge, which are quite common in certain districts. In the ravines between the Russian river and Smith's valley, which contain much shrubbery, they are constantly met with. They resemble their eastern congeners except in color, which is that of slate.

Next in point of numbers is the dove, of the same species as that east of the Mississippi river.

Ravens, as well as crows, are not very numerous, but sufficiently so for the few fields of maize in the country, and for the exclusive right to plunder for which they sometimes contend. The larger ravens, though, usually eject their smaller friends.

Vultures are occasionally seen.

Hawks, of course, hover in ample numbers where there is so much prey.

Near the marshes and larger streams there are vast numbers of curlews, as well in the Sacramento valley as near the seacoast.

The abundance of geese and ducks that frequent the rivers and bays in winter, has been referred to by several writers.

Of the goose, I was informed, there is one species which is probably identical with our wild goose, and another species of a white color. Both kinds, instead of remaining on the water, are generally found feeding on the fresh grass in the valleys and on the hills, little heeding the approach of man.

There are many varieties of wild ducks, but no particulars could be gathered in relation to them.

The seacoast abounds in birds of many species, among which is the penguin, in such vast numbers as to cover all the smaller islands with guano, in some instances to a considerable depth. They seem to take and retain exclusive possession of these small rocky islands.

Some writers have asserted that fish were abundant along the coast and in the waters of California; others; again, complain of their great scarcity. It should by no means be assumed that there are few kinds of denizens of the sea on that coast, because, as yet, only a limited variety is known. When I first arrived at San Francisco it was asserted by many persons that there were few fish of any kind in the bay, and none worth the taking, but before I left the country they were taking several kinds with the hook for the markets; which, some persons discovered, was more profitable than digging for gold. The natives, not being able to use the lasso in such operations, had never troubled themselves about fish; and in supposing the Yankees were in some way connected with their advent to the waters in the vicinity of San Francisco were not far wrong, because it is to be expected that the spoiled provisions and garbage from the city, and the three hundred vessels in the harbor, will *concentrate* certain kinds of fish therein, as well as in the strait which connects it with the ocean.

There is an excellent fish for the table caught with the hook outside the straits, as well as off Bodega bay. They have large scales of a dingy red color, and weigh several pounds. They are only taken at sea in not less than 25 or 30 fathoms of water *with rocky bottom*. Smelt are at times extremely numerous, and seem to feed upon the sardines with which the waters along the coast are often filled.

As the southern country can produce olive oil in any desired amount, it is probable that the sardine will hereafter become an article of export.

Halibuts of great size are occasionally drifted on the shores of Bodega bay, having been wounded by some other animal, probably the seal. I saw the remains of one which a couple of cuyotes were regaling themselves upon. I was informed that sometimes they are thrown upon the beach by the waves yet alive, but injured by the loss of a large mouthful of flesh.

The common hair seals occupy many places within the bays and inlets in numbers sufficient to constitute a branch of industry that will not be neglected after the auriferous excitement shall have subsided.

Whales having a *respite* are sufficiently numerous; among them is the enormous "California whale," as he is called, whose strength, activity, and *intelligence*, often enable him " to defeat his assailants," and compel them to retire with loss.

Sharks are not wanting, and, to *my* surprise, it was found that a couple of very small ones upon being cooked were by no means unpleasing to the taste.

The remains of a small lobster were observed upon the beach at Bodega bay, but no information in answer to my inquiries was elicited in reference to this animal.

There is a great variety of crabs, of which some are large and edible, and others extremely small. Many of them are beautifully colored upon the exterior.

The rocks in the bays and inlets, in some places, teem with a salt water crawfish, much resembling those of the small streams in the Atlantic States.

It has been stated that oysters were to be had in abundance from the waters of California, but I could not ascertain from any reliable source that this species of mollusca had been seen. The animal called haliotis is eaten to a considerable extent by the Indians, but no opportunity occurred to test its gastronomic properties. The haliotis·has an *open* univalve shell, which is sometimes a foot long and nine or ten inches broad, and being very irridiscent and prettily varied in color, is extensively used in the manufacture of many fancy articles in China, Europe, and America.

Large edible kinds of muscles are found, closely packed together and adhering to the rocks along the coast; the shells are of a purple color, and the mollusc itself is well flavored. It is possible the report about oysters may have arisen from the existence of either of these two last named shell-fish, both of which are extremely abundant.

Of the fish that inhabit or frequent the rivers and streams of the interior, the salmon has been so frequently described that it need only be remarked, that as soon as the waters rise in winter they crowd into them in myriads, especially into the streams which are discharged into the ocean north of San·Francisco bay, as well as emptying into this bay.

They become less abundant towards the southward; and during the spring, or early in summer, they return to the sea, and steer their course for the icy regions of the north.

Salmon trout are known to inhabit the Sacramento river, but are probably not abundant.

I was unable to ascertain whether the common trout existed in the constant streams of the mountains; none had been seen that I could hear of, although some trials had been made for them.

A large fresh·water fish, from one to two and a half feet long, with coarse gray scales on the back, and white beneath, is very plentiful in the fresh-water streams of the Sacramento valley, but is coarse and bony, so as to present little temptation to the epicure.

It is more than probable that other useful fish will hereafter be discovered in this country. Many valuable kinds may exist in its waters whose particular *habitat* is unknown; but the *investigating* propensity of our people will, before many years, discover at least all the means of deriving wealth from the sea, and its affluents, in California.

X. CLIMATE.

Before considering the agriculture of California, the subject of climate should claim attention; but the absence of sufficient meteorological observations causes great difficulties in the way of conveying an accurate view of the *many different* climates of this regoin; and from this has proceeded, in part, the misapprehensions which seem to have existed in the minds of most of those who hurried to that country within the past year.

Complaints are constantly made of incorrectness in the statements of those who have written upon this subject.

Most of our countrymen, it would seem, expected a universal prevalence of soft breezes and bright skies. They were sadly disappointed when they found on or near the coast strong northwest winds, carrying with them cold fogs and mists, which prevail during much of the year. In the winter—as the rainy season is improperly called—these fogs are less common; and in the intervals of days, and even weeks between the

Ex.—4

rains, which last for a few hours, (and sometimes in showers for two or three days at a time,) there is bright and pleasant weather. It is affirmed at San Francisco that the mean temperature is as high in winter as in summer; but accurate observations must be continued during some years before this can be fully credited.

The fogs and chilly winds of San Francisco, and the excessive heat in parts of the Sacramento valley, have beèn signalized by Frémont and others; and if those who were disappointed did not observe the notice, it was, perhaps, because they were too much under the influence of the *auriferous* excitement to observe other than the *bright* side of the picture.

As far as can be gathered from authentic publications, and much inquiry of the elder residents, besides personal observations during a few months, it would seem that there are few portions of the globe whose climate is not represented in California, if we include the peninsula.

During more than half the year, including the summer season, the cold northwest winds before referred to blow almost daily upon the line of coast from beyond the northern limits of California to Cape San-Lucas. Northward of Point Conception they strike the coast obliquely, whilst to the southward of it the coast trends to the eastward, so as to be sheltered in some measure by the spurs from the coast range, which run westward to this point. In its passage over the spurs the air loses much of its moisture, which, being condensed upon the hills, renders sensible an amount of latent heat sufficient to produce the marked difference in climate near the coast, which all agree in declaring to exist, north and south of Point Conception.

Near San Diego the coast line stretches more to the southward, so as to receive the northwesterly winds; but by the time they reach this low latitude their capacity to retain moisture in the state of *invisible* vapor is so much increased, that fogs are both light and unfrequent

These winds from a northern region, during their passage over thousands of miles of ocean, become supersaturated with moisture, a portion of which, to the great discomfort of navigators, is visible in the shape of fogs and mist which cover the ocean over a large area—at some seasons continuously for weeks, and at other times the sun dispels them during a portion of the day.

The heat-conducting property of air, bearing the relation to temperature and moisture which exist in this case, is very considerable, and thus produces the low temperature which otherwise could not be expected during the summer within this range of latitude.

The mild temperature of the winters would be looked for from the fact of the greater prevalence of southwesterly winds; and even a northwest wind would in some measure moderate the cold of winter, because of the higher temperature of the sea over that of the land, whilst the sun is in the opposite hemisphere.

The coast range of mountains receive the moisture from these northwest winds of the spring, summer, and first half of the autumn, during which they prevail to a great extent; and when they have crossed the mountains and reached the Sacramento valley the amount of sensible heat produced, in connexion with the diminished conducting power of the air in regard to heat, contribute to produce a climate widely differing from that nearer the coast; being, instead of cold and foggy, exceedingly dry and hot during more than half the year.

East of the coast range the sun is rarely obscured by clouds during the dry season. During thirty days from the fourth of July that our party was in the valley and upon the flank of the Sierra Nevada, the sun was only obscured during one afternoon and the following morning. With this exception, we had a clear bright sunshine during every day.

Passing from the valley, the westerly winds, deprived of their moisture to a great extent, roll upwards to the summit of the Sierra Nevada, whose reduced temperature, owing to the great elevation of its peaks, is sufficient to condense almost their last atoms of water, and the air passes into the "Great Basin" nearly in the state of absolute dryness, and thus contributes largely to produce the warm and arid climate of that region.

The deposites of dew are very heavy near the coast, but lessen further west, after the highest ridges of the coast range are passed. They are also at times copious in the valley of the Sacramento near the rivers and tulé marshes, but disappear in receding from the valley. We often found an entire absence of dew whilst upon the elevated portions of the mountain's flank.

At San Francisco the temperature of the air rarely reaches as high as eighty degrees; but instead of that equable climate that many looked for, the maximum and minimum within the twenty-four hours often differ as much as 30°. During the rainy season the temperature seldom falls below 40°, and snow is so rare an occurrence near the coast, that when enough fell upon one occasion, during the winter of 1848-'49, to whiten the hill tops, the superstitions of some of the older residents attributed it to the coming of the "Yankees."

There is a marked difference in climate after the first ridge of the coast range is passed, even within the distance of a few miles. This is exemplified in Smith's valley, which is separated from the bay of Bodega and the ocean by a ridge between three and four miles broad, and of no great altitude, and yet the climate is very sensibly drier and warmer than nearer the coast or even at San Francisco, which is further to the southward and at a greater distance from the ocean, but is amply supplied with fogs, through the straits which connect the bay with the sea.

The many valleys east of the more elevated ridges possess a most delightful climate. Among these are the Pájaro and San José south of San Francisco bay, and on the north the valleys of Petaloma, Sonoma, Nappa and Suisun, which are each sheltered in different degrees from the chilling effects of the westerly winds by the causes before adverted to.

Benicia is about twenty-five miles in a direct line from the ocean, and in a position which can only receive the northwesterly winds after they have been deprived of a portion of their moisture in passing over the elevated ridges westward of the bay of San Pablo. They blow, however, with considerable force through the straits of Carquinos into Suisun bay. Although fogs are often seen upon the summits of the highlands in the vicinity of Benicia, they do not usually extend so low as the site of the town. It possesses, in fact, a dry climate; and although the maximum temperature is usually not excessive, yet there are occasionally very hot days. On one occasion, (about the 20th of June,) the thermometer indicated 105°, at the same time that it stood at 80° at San Francisco, not thirty miles distant. However great the temperature may be during the day, the nights are pleasantly cool.

From every point on the flank of the Sierra Nevada whence a view can be had in a westerly direction, there appeared at all times a very hazy atmosphere over the coast range. But from the latter, the crowning peaks of the Sierra, at the distance of over one hundred miles, may be seen most distinctly defined. There was no hazy appearance upon any part of the Sierra that I saw, except in the vicinity of burning forests.

The Sacramento valley, especially near its western borders, is subject to great extremes of temperature. In winter the thermometer falls below the freezing point, and in summer it indicates, at times, a temperature of 110° to 112° at Suter's fort. The diurnal variation is also sometimes very considerable. On the 30th of July, at our camp, near where the Calaveras river leaves the hills, the thermometer was 91° at 3 p. m., and at 48° at sunrise next morning.

We found also a very high temperature upon the slope of the Sierra— generally from 85° to 100°—upon the hills in the afternoon, and were informed that a few days before we reached Coloma, (on the 17th of July,) it was 110°.

We found on the highlands, four miles west of that place, at 5 p. m., the temperature to be 93°.

It is quite probable that it was considerably above 100° (in the *shade*, as in the case of all the preceding observations) in many of the ravines which we were in, that contain but few trees. They reflected the sun's rays most powerfully.

There is one peculiarity in the climate of this mountain region deserving of notice; which is, that this excessive heat does *not* produce here the effects upon the animal organism that similar temperatures do in many other regions.

In the "States," when the temperature approaches 100° we hear of out-door labor being suspended; of sun strokes, &c.; and as lesser effects, there succeeds to the excited condition of the animal system, produced by exposure to great heat, a state of great relaxation. These lesser effects were perceived in a very slight degree, and no injuries of a more serious nature were heard of. It appeared to me, after a day's ride under the bright rays of the sun on the flanks of the Sierra, when the thermometer in the shade indicated a temperature of 100°, that the effects upon the system were not greater than would have resulted from remaining exposed to the temperature of 85° within the tropics, or even in the United States.

On endeavoring to get at the cause of this difference, the only one that seems sufficient is, that the *extreme* dryness of the air produces a rapid evaporation of fluids from the surface of the body, and that the heat which is absorbed by these fluids, and rendered latent as the evaporation goes on, is sufficient to leave the surface of the body as cool as it would be in an atmosphere of much lower temperature with a greater amount of moisture. It is certain that the evaporation of the perspired fluids from the surface proceeds very rapidly in these regions with persons in good health, but those laboring under high fever must suffer exceedingly in such a heated atmosphere.

The previous remarks were intended mainly to give an idea of the climate between the ocean and the summit of the Sierra Nevada, and between 37° and 39° north latitude during the dry season; and it remains to speak of the rainy season, usually termed winter.

There is really no weather in California to which the term *winter* is applicable, (in the sense we usually apply the word,) except upon the summits and elevated portions of the flanks of the Sierra, and upon other high mountains in the northern part of the territory.

The rains begin in the north, early in the autumn, and, extending slowly southward, usually reach the bay of San Francisco about or before the first of December, and San Diego a month later. They extend still further south into Lower California about to latitude 28° 30' north, where they seem to terminate. Southward of this point there are about 3° of latitude in which there is rarely one shower a year.

The country still further south, to Cape San Lucas, receives the showers from the northern edges of the tropical rains, which progress northwardly from about the equator, late in May, and reach Lower California in July, giving showers occasionally for a couple of months, and then recede again to the southward.

The rains from the north become very slight towards their southern limits, and the showers continue during a brief period, after which they cease in the inverse order of their progression when they commenced. The rainy season ceases at San Diego usually by the last of February, and at San Francisco in April, and so on northward. It not only lasts longer in the north, but more falls within a given time.

At San Diego the showers are mostly slight, and there have been seasons when scarcely any rain fell at all.

The temperature of the country between this place and Santa Barbara is much more uniform than that further north. The gentlemen attached to the boundary commission informed me, at San Diego, that they had experienced no weather during the summer that could be deemed in the least oppressive; and in the winter season there are occasional showers, with intervening spells of bright, pleasant weather. At this season vegetation makes amends for the period of drought, when nearly every herbaceous plant except the cactus ceases to grow, unless water be applied by artificial means.

In those parts of California that I visited, there was not the occurrence of a single night that was uncomfortably hot. However hot the day, the temperature at night was not so high as to prevent sleep, whether among the Coast Range mountains, in the Sierra Nevada, Sacramento valley, or at San Diego.

XI. AGRICULTURE.

With very few exceptions, the valleys and even the ridges of the coast range are covered with a soil which appears to contain in ample proportions every material requisite for a high degree of fertility, wherever the necessary amount of water is supplied.

From that part of Lower California wherein it rarely ever rains, we find the amount of vegetable growth increases with the increasing amount of rains to the northward. This was very evident, even as seen from the steamer, in passing up the coast during the latter part of May.

The tropical rains had not then reached the southern portion of the peninsula; and as the ship usually kept very near these bold shores, an opportunity was afforded to note, that, with very few exceptions, (occasioned by the existence of springs,) the western slopes of these highlands were entirely destitute of vegetation south of the 30th degree of north latitude. It is known, however, that in the interior of Lower California there is a number of small ravines and valleys, highly productive in consequence of the springs of water they contain.

In approaching San Diego, green hillsides dotted with cattle were observed after passing latitude 30°. Trees also began to appear of small size. But agricultural operations are out of question, except where irrigation is practicable. There are valleys among the lower hills about San Diego containing springs, with the aid of which various fruits and plants, both of the temperate and tropical regions, are cultivated. These watered valleys are more numerous in proceeding towards Point Conception, and immense numbers of cattle and horses are raised upon them and the adjoining hills.

Proceeding from the coast hills, (which are naked, except during the rainy season,) we are informed by Frémont and others, that upon reaching the more elevated ridges inland, they are found to be covered to some extent by trees and herbage during the entire year. This is to be expected from the reduced temperature upon these ridges condensing the moisture from the air.

Northward of Point Conception, the climate *along the coast* is unsuited to the habits of some of the fruits that grow so luxuriantly to the southward, as well as in the sheltered valleys between the interior ridges of the coast range as far north as the bay of San Francisco, and even in Sonoma, Nappa, and Suisun valleys.

The more copious supply of moisture during the rainy season northward of Point Conception, as well as the very heavy mists during the spring, occasion a better adaptation of this northern coast region to the cereal grains and grasses, and other agricultural productions of the temperate zones.

The ridges about San Francisco bay are too low to produce a sufficient amount of condensation to sustain, in summer, much other vegetation than trees in certain spots.

Northward of this bay, and that of San Pablo, we find the productiveness of the soil is increased, because moisture is more copiously supplied as a more northern latitude is attained ; until, near the Oregon boundary, rains fall at all seasons of the year.

The longer continuance of the rainy season, and the heavier rains, fertilize a considerable area of ground, and give rise to an ample number of constant springs, as appeared from the best information that could be procured.

There is an unaccountable phenomenon at Bodega bay that may be mentioned in connexion with springs. In describing the geological structure of that region, it was stated that "a narrow tongue of land extended from the northern side of the bay for three miles to the southward." This strip of land (see fig. 9) is about half a mile wide, and is bounded by the salt water of the ocean on the west, and the equally salt water of the small bay on the east ; it is separated from the hills on the north by a low narrow neck of land ; and is nearly now, what it probably once was in fact, an island. It seems remarkable that fresh water should occur under these circumstances, and yet there gushes from between the eastern edges of the

sedimentary rocks above the leptenite, (before noticed,) a number of copious springs, not in or near one spot, but distributed along the greater
portion of the side next the bay, towards which they necessarily flow, because of the strata dipping in that direction.

The quality of the water is equal to any, for its purity, I ever tasted.
The existence of these springs is the more remarkable, from the fact that,
for several miles in the interior, the country is nearly destitute of springs.

When treating of the structure of the country between the ocean and
the Sacramento valley, (the coast range,) it was stated that it consisted of
many ridges parallel, or nearly so, with each other ; and further, that
some of these ridges were composed of strata of sandstone and indurated
clay, dipping in opposite directions from the anticlinal axes at their summits, which strata were found, in some instances, to dip under intervening
valleys, and rise up again from their synclinal axes (see fig. 6.) These
trough-shaped strata present circumstances very favorable to the expectation of water rising to the surface, through artesian wells of no great depth.
This subject was fully explained whilst treating of the Sacramento valley.
If such trough or basin-shaped stratification exists in the arid country
south of San Francisco bay, it may alter the present prospects of such valleys, in their relations to agriculture, most materially.

The influence of stratification in determining the flow of water is a
matter of much interest in many parts of the world, but *nowhere* of more
than in California.

Towards the close of the tour with General Smith and suite, and after
Commodore Jones and myself had separated from the other gentlemen, an
opportunity occurred to refer the Commodore to a good illustration of the
above remark. We were passing through a series of small valleys, flanked by high ridges, between Armidor's reach and Benicia, when we noticed
that from the hills on the west side of us there issued many very fine
springs, some rising near the basis, and others at considerable elevations,
reaching to as much as 400 feet above the surface of the valley. The
parched hill-sides on the east were without a single green spot to indicate
the existence of water.

These " valleys of denudation" were situated between the hills, whose
strata are parallel with each other, as in fig. 2. The water flowing between
the seams of the strata was freely discharged in springs on the western
slope, (a) because the strata dip towards and crop out on that side. The
strata on the eastern side (b) not only prevent the discharge of water at the
surface from between them, but, during the rainy season, much that falls
upon the slopes *from* which the beds of rocks of these kinds *dip*, flows off
between the seams. Therefore the lands on the eastern side of the axis
of the valley must dry up much sooner after the rains cease than those on
the other side.

This is a subject which ought *not to be neglected* in adjusting a system
for laying out the public lands of this country for sale.

The extreme northern part of the Sacramento valley is within the
range of the summer showers, but to a less extent than the region nearer
the coast, because of the high mountains on the northwest.

The several accounts of its soil show that it varies from very tenacious
to very light; and considerable areas are much mixed up with sand and
some small gravel. Those portions nearest the San Joaquin and the Sacramento rivers consist of a deep and exceedingly rich loam, much of

which is annually overflowed; and besides these, is a very large area permanently covered with water of little depth, which produces the gigantic bulrush called *tulé* in California. These rushes are often ten or twelve feet high, and an inch in diameter at the base. On each side of the low grounds towards the hills the surface is more elevated, and the soil often less fertile, consisting of mixtures of sand and gravel, with a finer loamy earth; much of which, however, is sufficiently compact and fertile to produce very heavy crops of wheat. So far as my observations extended, there is little soil of this valley that would not be very productive in *some* useful crop, if amply supplied with water.

There is every probability that this may be effected (by means of artesian wells, as has already been pointed out) to an extent sufficient for the ordinary purposes of the farmer, as well as to irrigate his fallow crops—such as maize, potatoes, &c.—as well as all kinds of garden vegetables and fruits.

Most of the lower lands near the rivers do not require irrigation; but, such as do, can be amply supplied, at a trifling cost, through the agency of wind, acting upon machinery of a very cheap construction. There is scarcely a day in the year that there is not wind sufficient for such purposes. In fact, the scarcity of water-power and *absence* of coal are likely to bring this means of obtaining motive power into extensive use in California.

There are sufficient reasons for believing that the small number of springs in this region is more owing to the extreme dryness of the air than the scarcity of water within the reach of wells of moderate depth. We know at home that the water of many springs and wells fails during a long-continued drought; and as California experiences a drought of six to nine months annually, the same may be expected to occur there. In such cases the needed supply is usually to be had by deepening the well.

As but a small proportion of the country is thickly wooded, so as to be sheltered from the rays of the sun, the ground becomes heated to a considerable depth, and evaporates the water of at least all the small underground streams, as well as those which approach the surface at a very slight angle thereto, instead of a direct line from a considerable depth. In the Sacramento valley there are many wells from ten to fifteen feet deep which abound in water. The structure beneath this valley, and its position, would indicate that it is saturated with water below the level of its principal rivers.

Very shallow wells furnish water in some of the tertiary valleys of the coast range that I visited; but in the hills and among the hypogene rocks its attainment by means of wells would be uncertain in many places.

It is more than probable, that if the coast-range region generally were wooded, as is the case farther north, there would be no scarcity of water within its limits.

The agricultural capabilities of the western slopes of the Sierra Nevada are altogether unpromising southward of latitude 40° north. Northward of that line; it is believed, the country is sufficiently moist, not only upon the lower mountain slopes, but in the valleys. Southward of the parallel of 40° there is a belt of well-watered country parallel to the range from the base of the snow-peaks for some miles downwards, which, although not sufficiently attractive to induce emigrants to settle at present, will come into play hereafter, and aid in supplying food for the more arid region in which the gold occurs.

The remaining portion contains a very small area, in the aggregate, which possesses the least value in reference to agriculture. The few arable spots are made up of the small number of valleys, each of little extent, which contain springs or streams, that do not cease to flow in summer.

There is variety in the soil, but in general its appearance indicates a composition wanting in most of the elements of fertility; so that but little could be expected in return if irrigation were practicable and practised. This, however, cannot be done to any extent except near the western termination of the slope, in the vicinity of where the large constant streams enter the Sacramento valley, which most of them do, through beautiful small valleys of very fertile soil, most of which will produce heavy crops of grain without irrigation. These valleys generally extend several miles from the great valley, and throughout the year are clothed (where not cultivated) with a luxuriant growth of herbage. Their appearance is greatly improved by open groves of oak trees, whilst the rivers are fringed with the willow, a tree whose appearance is often looked for with the greatest interest by the traveller in California as indicating the existence of water.

If, however, water be not wanted for agricultural operations, there are other branches of industry for which it will be required, and that it may be had in limited quantities is very probable.

The *moisture* observed in the lowest places in *some* of the ravines and valleys among the highlands, (where we rarely found a running stream,) indicates that water might be obtained in many places from the ordinary kind of wells of moderate depths. This means of getting water will be the principal reliance for the purpose of washing gold in the highlands during two-thirds of the year, except on the borders of the greater streams. The geological structure of the region forbids the hope of succeeding by means of artesian wells, as often as in one case out of perhaps a hundred.

Except among comparatively few persons, mostly in the southern country, who raise grapes and other fruits, there is little culture of any kind. In fact, in agriculture most of the holders of those great bodies of land, with their large herds of cattle, have not, generally, advanced much beyond the state of this branch of industry in the days when Jacob was obliged to send his sons to Egypt to purchase grain. It is true, however, there are very few who have *begun* to think about the propriety of some improvement. They have numerous large herds of horned cattle roaming over the hills and valleys in the wild state; but the catching of them, and of the equally wild horses, constitutes the principal amusement of many of the lords of the ranchos and their dependants; and not a few content themselves with beef as their only diet for long periods, unless they chance to have a little maize to make tortillas.

The more enlightened cultivate gardens; but there are a great many who will dispense even with the onion and red pepper rather than trouble themselves with raising them, probably because they cannot attend to it on horseback. They seem unwilling to do any kind of work, unless it can be done in this equestrian fashion. I have noticed the drivers of horse-carts and ox-carts working in the streets of San Francisco mounted upon extra horses; but they did not long endure Yankee competition, which dispenses with a horse for the driver to ride.

The pastures produce very large cattle, whose beef is delicious. Milk, cream, or butter are extremely rare among the native rancheros, with often hundreds of milch cows oppressed with superabundance of milk; and few of them ever take the trouble to be supplied with tamed cows—they prefer usually to dispense with dairy products; but when they do take a fancy to have milk, a couple of vaqueros mount their horses, and course over hill and dale until they find a cow that gives promise of furnishing a good supply, when she is lassoed, dragged and driven to a post near the house, and made fast by the horns. Her hind legs and tail being then fastened together, one party performs the operation of milking, whilst the other holds the bucket.

The oxen are better than the best in the United States in their ability to perform severe labor. They are very large, well-formed animals, and, as well as the cows, would astonish our farmers if some of the best were exhibited at our eastern cattle-shows.

The horses are rather small and delicately formed; but if pains were taken in crossing the breeds, a superior race of these animals could be reared. Like their Barbary progenitors, they are capable of long-continued exertion without food or water.

There is but little of California that is not peculiarly adapted to raising sheep, and the only difficulty now in the way arises from the vast number of cuyotes all over the country, which renders it necessary to confine the sheep at night within an enclosure that cannot be scaled by these animals, or the large wolf which is occasionally seen. The sly movements of the cuyotes makes it necessary that sheep be watched whilst pasturing during the day.

Swine may be raised with great advantage in the many parts of the country where oak trees abound.

The success which has followed the efforts of Americans and others in raising chickens, proves the climate in many parts to be well adapted to this class of poultry, especially where the heavy dews (so destructive to poultry in the United States) do not prevail. I did not notice turkeys, but it might be expected they would succeed better even than chickens. The vicinity of the water-courses affords admirable situations for rearing the aquatic classes of poultry here, as elsewhere.

One would be prepared to expect, in a country with so many varieties of climate, an extensive range in its productive capacity in regard to plants; and such we find to be the case in so great a degree as to justify the remark, that there are few plants of value to man that cannot be raised in some part of Upper California, except such as can only grow in an extremely moist climate. But disregarding the greater number, we shall only refer to the most useful kinds that seem adapted to the country.

Heavy crops of *wheat* may be readily raised over a very large extent of country northward of Point Conception, embracing such portions of the Sacramento valley as are not overflowed in the winter, or which are not covered by sand and gravel, as well as upon the greater portion of the *hills* and valleys of the coast range. It is not unlikely, also, that the wheat-producing district may include the valleys among the high ridges still further south, unless, as is probable, it may prove more advantageous to apply that region to the pasturage of cattle for the use of the southern country.

The experience of the country justifies a very favorable opinion in reference to the facility with which wheat may be procured; and, in considering the habits of this important plant, it would seem that the climates within the range before recited are well adapted to it. With the first rains of the season the ground is so far softened as to be workable, when it is ploughed and roughly prepared, and the wheat sown as early as possible in December, after which it soon comes up.

The frequent rains, with intervening sunshine, and the absence of freezing weather, keeps the plant in a healthy condition, without growing *too* rapidly, until the latter part of February or March, when the brisk growing season commences; and, except in case of an unusually early termination of the rains in the spring, the growth and maturity of the plant is completed by the time the soil becomes dry. In the mean time the rains have ceased, the farmer feels sure of dry weather to secure his crop, and has no fear of rust, mildew, and many other causes of injury so often complained of among us. Winter-killing, or other effects of frost, cannot occur; and I could not learn that the fly or insects of any kind injured it, except that in the unusually dry season grasshoppers sometimes were troublesome in a few situations. The crop is mostly fit to be cut during the last two weeks in June; but is often allowed to remain much longer to suit the convenience of the farmer, and in doing so sustains infinitely less injury than in a climate where rains may occur at any time.

This dry feature in the climate also renders stacking and housing unnecessary until the return of the rainy season; yet a careful farmer would, if possible, cut his wheat as soon as it matured, and secure it from the depredations of various animals.

There is *much* to be improved upon in farm management, in this as in everything else, in the agriculture of the country.

The thrashing of wheat is performed in a very expeditious manner in this country. An enclosure is made large enough, in many cases, to hold the whole crop a couple of feet or so deep; and, after it is so disposed, a large herd of wild horses—often more than one hundred—being turned in, are followed by a proper attendance of mounted vaqueros, who, by dint of chasing, whipping, and shouting, keep the whole party in such rapid motion that there is little opportunity afforded for the animals to appropriate any of the grain to their own use.

It is not easy to ascertain the product of wheat upon a given area from California farmers; but the best information that could be obtained is, that 35 to 40 bushels can be readily raised to the acre in the present rude way of farming, and without manure. They know better the number of fanegas sown and gathered, which has in some instances amounted to one hundred for one.

What has been said in reference to the raising of wheat is also applicable to rye; and it is not improbable this crop might be grown in the upper portions of the Sierra Nevada; but being inferior in estimation, it is not likely to be extensively raised.

The range of oats is coextensive with that of wheat; in fact, a large area of country within the Sacramento valley, and westward thereof, is annually covered with wild oats; and in the richest lands it grows most luxuriantly, forming an important item of pasturage; and, for even months after it is entirely dry, cattle and horses will actually fatten upon it. That it should thus retain its nutritive properties after remaining so long exposed to the weather, can only be accounted for upon the supposition that the extreme dryness of the climate prevents all chemical action upon its azotic and other nutritive components, and that it remains on the ground precisely in the condition it would be if cut and secured.

It has been asserted by writers—but upon what authority I know not—that the extensive distribution of this grain results from the sowings of the early colonists, and that it has subsequently spread over the country. This requires confirmation, because the grain is smaller and otherwise different from all the varieties of cultivated oats which we are accustomed to see, and, although there is apparently little substance in the grain, the entire plant must, from the results produced, be highly nutritious.

The manner in which it is annually sown and the seed covered in by natural causes, is worthy of notice. During the dry season most of the soils throughout the country which are sufficiently tenacious to be productive are traversed by deep cracks and fissures running in every direction, and so numerous as rarely to leave a space between them one foot in diameter.

In some low situations, where the soil is very deep and stiff, the fissures open to a width greater than that of a horse's foot, and thus renders travelling over such ground somewhat unsafe for horse and rider.

As the matured seeds of the oats fall to the ground, a portion of them enter these openings in the dried soil, where they are as secure from injury and depredation, until the returning period of vegetation, as they would have been in the stack or barn.

This matter was investigated by Commodore Jones and myself, whilst travelling together, upon our return from the reconnoissance before referred to. The Commodore, being alike successful whether in ploughing the waves or the soil, takes a deep interest in matters relating to agriculture.

We observed that the stalks of oats were not standing uniformly over the surface, but in *lines* crossing one another, and running in every direction like the cracks in the soil, but not coincident with those produced since the maturity of the standing crop.

The seed being deposited in these openings, remains *unaltered* until the advent of the rainy season, when the expansion of the soil closes the cracks and covers the seeds. During their period of vegetation, the roots bind the soil together, so that the openings for the next succeeding crop are not in the same lines. The whole operation is, in fact, a case of natural "drill husbandry" on a grand scale, and its occurrence may give *useful lessons in labor-saving* to an observing Californian farmer.

In sowing oats for the grain, it is usually put in the ground about the beginning of March. Whether it would not be better to sow it upon the cracked surface after the first showers, and harrow it in, is worthy of consideration.

No opportunity occurred to notice barley whilst in growth, but reliable information was obtained to the effect that it is well adapted to the country. The habitudes of this cereal clearly indicate that it is peculiarly suited to the circumstances of climate and soil which exist over a large portion of California. It delights in a dry soil—rich, light and loamy—of which there is an ample area in the coast range, and also on many portions of the Sacramento valley. It grows so rapidly, that throughout much the larger portion of the country it may be harvested during the latter part of June, or within ninety days of the time it was sown.

Loudon (Encyclopedia of Agriculture) states that "it has been known to grow and produce well without having enjoyed a single shower of rain from the time of sowing until it was harvested."

From the peculiar adaptation of this grain to California, it may be expected to supply the place, in a great measure, both of maize and oats.

It is only in very limited portions of the country that maize can be grown without irrigation. These are near some of the water-courses in the Sacramento valley, where there are narrow strips of very rich loam not much elevated above the water-level. Such corn-producing spots also occur in Suisun and other valleys west of it. Besides these, the information obtained gives reason to believe there is a considerable extent of land northward of the latitude of Bodega bay, sufficiently moist during the summer to produce Indian corn; but nearest the coast, where moisture is most abundant, the temperature is too low for large crops of this grain. At Bodega a small kind of corn was growing, fit for the table in August and September, upon which no rain had fallen since early in May.

Buckwheat had not been raised, so far as I was informed, but would succeed, without doubt, in many situations, especially in the more elevated moist regions, and in the northwest.

It is only in situations permanently moist, or where there are summer showers, that the useful perennial grasses can thrive, and of such there are many indigenous kinds; but whether any others could advantageously supplant them, there were no means of acertaining.

The plant which abounds in portions of the southern country, and is called "burr clover," is much liked by graminivorous animals. The only specimens that came under my notice did not *appear* to be leguminous plants, and probably not clover at all, but the specimens were not sufficiently perfect to allow a decided opinion.

Another plant, called "pin-grass," from the shape of its seed vessel, but which is not a grass, is much relished by horses and cattle, and enjoys the advantage of living through the summer in comparatively dry situations.

The rich soils in most of the numerous valleys are favorable to the growth of certain kinds of leguminous plants, which are greedily eaten by animals. Some of these "wild peas" resemble the vetch in appearance.

The common garden pea does remarkably well in the northwest, without irrigation. A curious appearance was pointed out at Bodega, which is unusual with this kind of pea. Capt. Smith had planted his peas early in the spring, as usual, and enjoyed the advantage of an abundant crop for the table early in June; after a time, and as the vines dried up, new shoots started up from the crown of the roots, and in their turn grew and produced a copious second crop, of which I partook in September.

The frijole, so often praised as a table vegetable in California, is very similar to the ordinary "marrow-fat" kidney bean, and requires irrigation in most of the country.

There is but a small proportion of California in which the biennial and perennial clovers will be likely to be grown to advantage. The roots would die during the long dry season. There is, however, a species of *annual* clover, cultivated extensively in the warm dry climate of the south of France, that deserves notice in this connexion. It is there sown either in the autumn or spring. When sown in autumn it affords good pasturage during the winter, or, if left till May, produces a very heavy crop of herbage. It bears various names: as "flesh-colored clover," "farouche," "trefle de roussillon," &c., but its systematic appellation is "trifolium incarnatum."

It is believed that any desirable amount of rice may be produced in California. There are extensive marshes adapted to this grain among the low lands within the principal valley, as well as in some of the smaller valleys which open into the bay of San Francisco, and the smaller bays connected with it.

The potatoe can only be raised in the lower levels of the southern country by irrigating the soil. It is very probable that the higher ridges contain favorable situations for this valuable root. The northwestern parts are remarkably well adapted to its cultivation. The potatoes of the region inland from Bodega are unsurpassed in quality. So scarce was this tuber at San Francisco a few months since, that a German settler was offered, in my presence, $3,000 for the produce of what was estimated at about six hundred bushels upon two acres of land; but, from the great rise in price soon after, they probably produced a much larger sum.

Turnips grow luxuriantly through the winter and spring everywhere. Beets, parsnips, and carrots are also readily raised, and in very many places without the aid of irrigation.

In those situations which have already been noticed, where water can be obtained for the purpose of irrigating the soil, almost any plant may be raised. Besides all the necessaries and luxuries of life that are produced in Europe and America, there are few plants of the tropical regions that cannot be grown in the southern country and in the low interior valleys as far north as Nappa, and where figs, dates, sugar-cane, and even bananas flourish.

Apples and pears are found to do remarkably well in the north, as the orchards planted by the Russians at their farming establishment at Ross testify. Peaches were raised at all the missions; but as that fruit requires for the protection of the tree against insects, and for the full development of its luscious flavor, a climate cold in winter with a protracted hot summer, it will best succeed among the foot-hills of the Sierra Nevada.

A very large proportion of the country is found to be peculiarly adapted to the growth and perfection of the grape, from which rich wines are made; and there is little doubt that after this branch of industry shall be efficiently prosecuted, other varieties of grapes introduced, and the best positions for the vineyards ascertained, California will be celebrated both for the qualities and the quantity of its wines. The dry autumn of the lower country will give great advantage in the production of raisins and other dried fruits.

Among fruits that can be produced for export or consumption at home, are figs, olives, dates and prunes. The caper shrub would thrive in that climate, without doubt.

Although the coffee tree will grow in many situations south of Point Conception, it is not likely to be reared to much extent, because the ground would be more profitable if occupied with grape vines, &c.

It has been suggested that tea might become a product of California; but it may be doubted whether it would succeed so well in any part of that country as near our eastern borders, upon the hills of the Atlantic slopes south of the Chesapeake bay, where the climate is similar to the tea-growing districts of China.

It is unnecessary, in a mere sketch of the products of California, to enumerate other than the most important articles that are or may be produced by the soil. Some of those named can be raised in very large amounts, both for home consumption and export; others will contribute to furnish supplies of necessity as well as luxury for the use of its inhabitants. To those for home use might be added nearly all the esculent vegetables and fruits of the tropics and temperate regions of the globe.

In the present state of agriculture, the subject of amending the soil by the addition of manures is not thought of, nor will attention be turned thereto, so long as there remains unoccupied such an extent of rich virgin soil. It is not, therefore, necessary in this memoir to say much in relation to the sources of manure that may be in store for future events

Enough was observed upon this subject to give assurance that ample means for amendment may be readily had, whenever the people will use them.

Kelp, as was before remarked, is exceedingly abundant along the whole coast, and it will amply repay the cost of applying it to the soils near the navigable waters, for several reasons. In addition to the fertilizing effects due to its organic components, as well as mineral salts, it abstracts moisture from the air in a remarkable degree, owing to the very deliquescent properties of two salts that it contains, the chloride of calium and magnesia. This manure is so peculiarly adapted to dry soils, that it cannot be too soon brought into use in California.

The many small islands along the coast and in the bays are covered with guano, deposited by the myriads of sea-fowl which frequent them.

The quality of this material must be better and better to the southward, and it cannot be doubted but the guano on the coast of lower California is at least equal to that of Peru.

That marl will be found in many places in the coast range, there is scarcely a doubt, and, with shells from the seashore, will furnish the requisite calcareous matters. Those who may wish to use swamp "muck," will find, in some parts of the country, more of it than is at all desirable, both on account of the fevers these marshes produce, and the clouds of mosquitoes that leave them, to worry nearly all animated nature.

It is now less than two years since the only means of moment which California possessed for the purchase of imports consisted of hides and tallow. To procure these there were annually slaughtered great numbers of the finest cattle, and yet the progressive increase went on at a most rapid rate.

The population of the country being trebled in one year produced no scarcity of beef; and it was believed by the elder residents, whose opinions are to be relied on, that the consumption does not yet equal the annual increase. The enhanced price of beef is due to the great value of the labor required to furnish the meat to the purchaser.

The sudden demand for horses was also met by an ample supply.

Portions of many of the valleys are literally whitened with the bones of these animals, which have been slaughtered, or died in the course of nature. Whenever systematic and judicious farming shall be practised, these will form no small resource for improving the land.

In many places, there are dry spots of considerable area, apparently much frequented by cattle and elk, although nothing will grow there—which appeared strange, because the soil seemed richer than elsewhere. I regret that it did not occur to me at the time to procure samples of soil from these spots, which would probably have shown a large proportion of saltpetre, attracting the animals whose droppings, *in this climate*, annually add to its quantity. This subject deserves investigation, which may demonstrate this material to be worthy of attention hereafter.

It has been already observed that there is no timber of importance in the Sacramento valley, except the straggling oaks along some of the streams of water; and many of the valleys south of latitude 38° in the coast range are similarly situated in this respect, although there is a considerable number of them at no great distance from redwood and pine forests.

The means by which these lands, whenever they may be divided, can be enclosed, is a matter of some interest. At present, they effect the object by cutting ditches of such dimensions as will prevent horses or other animals that may injure the crops from leaping over them. Owing to the absence of frosts, the sides of these ditches maintained their nearly perpendicular slopes remarkably well, whenever they were observed, which was in every case in rather tenacious wheat land. How they would answer in less tenacious soils—where, also, there were not so many grass roots—there were no means of knowing.

In the dry regions of the mountains several varieties of shrubs were noticed, that seemed well suited for making hedges; and it may be found that a smaller ditch, with a hedge of some of these plants, will, in that climate, furnish a cheap and durable means for enclosing or dividing farms.

XII. PUBLIC LANDS.

The ordinary modes of surveying public lands, and dividing them, for sale, as practised by the government of the United States, are wholly inapplicable to the *peculiar* circumstances existing in California, whether in reference to the mineral lands or those applicable to agricultural purposes.

To lay off these lands in the usual quarter sections, with dividing lines parallel to the lines of latitude and meridians, in all cases, would not only lessen very materially the amount produced to the public treasury, but would give boundaries, in most cases, extremely inconvenient to the future occupants.

In surveying those not supposed to be metalliferous, (after sufficient geological examinations shall have been made,) there are matters for consideration not to be disregarded in any system which has for its object the location of these lands in such manner as will most conduce to the benefit of the nation at large, as well as those who may hereafter occupy them for agricultural or other purposes.

The point of *first* importance in California, in reference to lands, relates to their situation as regards water for ordinary farming purposes, as well as for irrigation in most of the country. This is a very simple matter in the smaller valleys, or parts of valleys, through which there are running streams. Narrow lots might be located from the water, on each side, to or over the hills, according to circumstances.

Towards the north there are many springs, constant themselves, but whose brooks in summer become dry often to within a short distance of the spring. Most of the lots in such districts would be valueless to the government, unless due care be taken, having reference to the position of the springs.

In those kinds of valleys where the stratification is such that all the springs are on one side of the valley, (fig. 7,) or on its bordering ridge, a different plan would become necessary. The farming lands should, in such cases, run *across* the valley, and extend over one or both the hills, according to circumstances.

In those districts where there are neither springs nor water-courses within considerable areas, much land will be nearly worthless, whenever the watered regions around them may be enclosed, unless it be ascertained that water can be supplied by artificial means for ordinary farming purposes, as well as for irrigating at least a small proportion of each lot of land. In some such districts, an investigation of their geological structure may indicate whether, by means of artesian or common wells, an expectation of procuring water may be entertained.

A very large area of lands, in the arable parts of California, must be laid off in lots upon which there will be no forest trees whatever. The inconvenience attending this state of things might be at least partially remedied, by dividing the nearest woodlands into lots of small size, so that each purchaser of farming land might be enabled to secure a small piece of forest land for the uses of his farm.

Although there are districts within which the lots might be laid down in squares, or quarter sections, of 160 acres each, (as in *portions* of the Sacramento valley,) yet, throughout most of the country, this system would be highly prejudicial to all interests, except to those of an occasional speculator or other acute person, who, by securing certain lots containing all the water and wood for a great distance, might enable him to prevent the sale of the remaining lands over a widespread area, so that his own lands would have the benefit of them in perpetuity, as fully as though he was lord of a " ranch " of tens of thousands of acres.

Where there are districts of large extent with few springs, at considerable distances from each other, and wherein the structure forbids the hope of water, there is no other resource but to lay out the land in larger lots for grazing farms, so that each lot should have water at least in one place.

The mineral districts will require the aid of science and good judgment to a much greater extent even than the agricultural portions of the country; but it would take up altogether too much space in this memoir to do more than throw out a few *brief* hints upon this branch of the subject.

Quarter sections, or other *square* divisions of small size, would be wholly wrong, because those situated near the lower portions of the slopes of the large ravines, upon which productive veins may crop out, would fall into the hands of comparatively few persons, who, owning the only *practicable* means of access to the valuable veins, would put it out of the power of others to work them; and thus they would remain unsold until the parties owning the front lots might choose to take them at the legal minimum price.

If, on the other hand, they were laid out in *very* large bodies, (each of which must reach somewhere a *deep* ravine,) the cost would be such as to prevent all except parties with considerable capital from purchasing. This, in some parts of the world, would not constitute an objection; but it is different with us, because inconsistent with the maxim of "the greatest benefit to the greatest number."

There are but two courses to pursue in the matter. The first will be, to make no mere *land* surveys at all of the mineral districts, but to have correct topographical maps of the country, made from actual surveys, for that purpose, and permit any one who may choose to work a vein to take possession of it, with a reasonable width of land upon each side of the opening to the vein, upon the payment of a *fixed* sum, with or without an annual payment of a portion of the proceeds as rent—which, however, in the cases of gold and silver, would no doubt be better arranged through the mint. The party might be secured against another party operating subsequently in the same vein, within a specified distance of the mouth of his adit, and be permitted to hold the right to work to said distance.

Whether this mode, which must tend to develop the metalliferous re-sources of the country rather tardily, would prove more advantageous to the public than one by which they can more rapidly be made *fully* available, not being a question of *physics*, is out of place here.

Another system may be briefly adverted to, as a means by which these resources may be developed in the shortest time, and which would seem to promise a larger addition to the public purse, whilst it could have the advantage of relieving the general government from the care of the property at the earliest period.

This system would commence by causing a minute geological and topo-graphical survey to be made of all the metalliferous districts. It is ne-cessary not only to be able to lay down on the maps the exact lines of the streams of water, and other mere geographical features, but the topogra-phy of the country must be accurately known by determining the heights and positions of the hills, valleys, ravines, and streams, so that the re-quisite sections and maps can be prepared. Whilst this is going on, the geological part of the work would be in progress, and make known the structure and composition of the rocks.

The veins which *abound* throughout the gold region would require a most detailed and careful investigation, and the surveys should be so con-ducted as to develop not only the more *obvious* veins of quartz contain-ing gold, but the different *systems* of veins, and to determine their metal-liferous characters; and also, whether there may be imbedded in the rocks masses of metallic or other valuable minerals.

The directions and dips of those *systems* of veins which may be found to contain matters of value, being ascertained, the next step is to deter-mine their position on the ground, (including the angle of dip,) and ex-act direction of the *productive* veins, whose thickness should also be de-termined, wherever it be readily practicable. It will only be *after* such a survey shall have been made that the metalliferous districts can be divided for sale, or permanent lease, with advantage to the public treasury, or to the many who may undertake mining operations.

Upon the acquisition of such knowledge of the region as has been in-dicated, the boundary lines of lots may be laid down to the best advan-tage. Each lot should contain one or more productive veins (see fig. 8) and commencing at the axis of a ravine, or at the water, if any there be, and with a proper width should run over the hills until it is arrested by a suitable line of division, separating the ranges of lots which are laid off from the nearest other ravines. The lots would generally have greater length than breadth, and their longer sides would be parallel to the more important productive veins, and to each other. A maximum, or constant width, might be fixed on by law, but their length would depend upon the distances between the several adjoining ravines, because there would be little wisdom in laying out mineral lots in this region upon highlands exclusively, or in positions where there are no productive veins.

By this system there would be lots of varied length and positions, some of which might be disposed of at high prices, and others so low that almost any one could secure a mining lot who was disposed to adventure in pursuits of this nature.

There are serious objections to granting leases (without certain reservations) for more than a year or two at a time, for the purpose of working the gold in the drift of the ravines.

The veins will be found cropping out upon the faces of these ravines and extending beneath them; and, however valuable a vein might be found, if the strip of ground between the foot of the hill and the water was, for some distance above and below, in legal possession of other occupants or owners, such veins could only be worked with their consent, unless at a greatly increased cost, by sinking shafts, pumping out water, and hoisting out the mined matter. On the other hand, if the flat pieces of ground where they exist, at the foot of the hills, be at the disposal of the lessees or owners of the *mines*, they will have not only room for the disposal of the matters excavated, but the water will flow out without cost, and the other matter be brought out at a far cheaper rate than by elevating them.

Besides, it is necessary to have room as near the mouth of the mine as possible, for performing the operations of sorting out the materials as well as grinding, amalgamating, and other necessary processes.

Water is absolutely necessary for the washing process; and water-power for grinding, pumping, rocking, &c., is very important in saving cost, which will be found generally to approximate nearer the amount of the receipt than is to be desired.

It would appear, therefore, that no leases of these " placers," as they are called, should be granted without the reservations which the circumstances render necessary.

It is not easy to realize the importance, in its fullest extent, of the caution proper to be observed in laying down a system for surveying and disposing of the public lands in general, in California, without a careful study of the subject in connexion with the applications that may be made of the lands in different districts. The *peculiar local circumstances* must be well understood, and be taken into consideration, both in the metalliferous and arable districts; and they are different altogether from any the government has hitherto had in charge.

It is *highly necessary* that certain reservations be made in reference to roads in the *deep* narrow ravines of the metalliferous districts, as otherwise there are places where individual owners might throw serious difficulties in the way of such improvements, required by the general interest. But the explanations in reference to this matter will be pretermitted, as this chapter already has exceeded its alloted limits. It may be added, however, that if the plan of surveying last indicated be adopted, roads may be *located* at the same time—at least in the deep ravines—so as to interfere (whenever they may be constructed) as lit$^{\text{tle as}}$ possible with the operations of mining, whilst they would facilitate the means of transportation.

XIII. CONCLUDING REMARKS.

In the whole range of its physical geography, California differs so entirely from the country east of the Rocky mountains, as to present a great difficulty in the way of comprehending its actual physical condition, (so far as climate and agricultural capabilities are concerned;) and this is utterly impossible, unless he who studies it is able to throw aside all ideas of comparison and approach it as a case *sui generis*. There is a hazard of committing mistakes analogous to those of certain travellers in the United States, who are unable to comprehend our social institutions, simply because everything appears defective or vicious that differs from what had come within their own sphere of observation at home.

The difficulties above referred to will not be removed for some years, and, in fact, not until such means be taken as will inform us accurately about the various matters upon which the physical constitution of the country depends, comprising a knowledge of its geology and topography, &c., from actual survey; and, in addition, there are required carefully recorded meteorological facts for several years at many points.

That it possesses abundant resources, wholly irrespective of its precious metals, there is ample testimony.

The cereal grains, except maize, can be produced to an amount sufficient for a very large population and to spare, without interfering with the rearing of animals to any desired extent; and this, too, whilst there will be left space for the vine, olive, fig, and a host of other valuable products, sufficient not only for home consumption, but for a large export trade.

The metalliferous prospects of the country have been noticed in these sketches; and to what extent its mines may be permanently productive cannot be estimated, nor can the more important one be arrived at for some time to come, of whether they will produce a sum proportionate to the amount of labor that may be applied to them; or, in other words, whether the same amount of labor applied to other branches of industry suited to the soil and circumstances would add a greater amount of wealth to the country.

According to the opinions of the best informed persons (whose residence in countries where most of the precious metals are produced, gave them excellent opportunities of forming correct opinions) the affirmative is too generally the case.

Whatever may be the practice among the stationary races of mankind in these respects, it is not to be expected that our people will persevere in a "losing business."

We are prone to look after varieties of pursuits, and will be likely, in general, to apply our labor to those which are most productive, notwithstanding the allurements presented by the appearance of gold and silver.

It is no less curious than true, that many persons in countries producing precious metals persist in working among them, often, when they cease to afford a remuneration equal to that of other pursuits carried on in their respective districts.

Many of our citizens hastened to California during the past year in

·consequence of the numerous exaggerated, one-sided stories, which were circulated in reference to the facility with which gold could be gathered. They had been told of various individuals who had collected large sums; and a *few* had done so; but the experience of the *many*, who did not pay expenses by gold digging alone, from the nature of the case, is far less likely to be known.

As with *lotteries*, the *few* who draw large prizes become subjects of conversation; but nothing is heard of the *many* who draw blanks, or prizes too small to pay the cost of the tickets.

In most cases, those in search of gold landed at Sán Francisco, and proceeded by water to Sacramento city, or Stockton, or Vernon, and from thence to the gold region; and the unlucky portion, after "*prospecting*" (Angl., searching for good diggings) over the hills and ravines of the Sierra for some time, with one single idea in view, even if their health remained unimpaired, became soured and disgusted, and, wishing the whole country as far under the sea as it ever was before, either write home in a gloomy mood, or return to San Francisco by the route they went, and thence home, if their health and means would permit. Even if they had travelled sufficiently, and were ordinarily capable of judging, such persons are not likely to be in a mood suited to look into details, sufficiently to form correct notions of a country where their high expectations, upon the only point that interested them, have not been realized.

Besides, they have rarely visited any of the arable districts, unless they may, possibly, have seen one or more of the small valleys opening into the western side of the Sacramento valley; and yet they cry, " 'Tis all barren." They have traversed, in the heat of summer, the dry hills of the Sierra, and in approaching and leaving them they passed a few miles over the dryest parts of the Sacramento valley; and yet *such* have ·often been heard to assert, there was "no land that could be cultivated in *all* California."

The principal productive resources of the country, it would appear, will consist of the precious metals, the cereal grains, the products of the vine and the olive, in addition to various domestic animals, among which horned cattle, sheep, swine, horses and mules are the most important.

There is one important product omitted in its proper place, which may be referred to at this time; it is *common salt*, which possesses the more interest from the expectation of a considerable amount being hereafter required in a country so well adapted to producing the requisite meats for preparing salted provisions. There are few climates in the world better adapted to the manufacture of salt from sea-water than is to be found south of Point Conception; and there is no question but it may be hereafter prepared to any extent upon many suitable locations, where, for at least eight months in the year, the process may go on without interruption from rains. The salt thus produced by slow *natural* evaporation, without artificial heat, is in large crystals, and being the purest kind, is preferred universally for curing meats and fish. A large proportion of the coast of lower California; also, is well adapted to salt manufacture, but whether convenient locations abound or not is not known.

After the present morbid state of things shall have passed away, each of these branches of industry will begin to assume its *proper* place in the general industrial system, and California will have its part to act in the trade of the world. With all its resources and an industrious population, it will not be likely to want the means to pay for supplies that may be more advantageously purchased abroad than produced at home.

Manufacturing, in the sense we usually understand the word, is not likely ever to be prosecuted there to any extent, because, among other reasons, of the small amount of water-power and the scarcity of fuel in a great part of the country. The steady winds throughout this region is favorable to the use of the power to be derived therefrom, and which will answer for all the ordinary wants of a country, including the manufacture of flour and other articles.

On the eve of drawing this memoir to a close, the newspapers bring accounts from California which go to furnish additional evidence of the correctness of views hereinbefore expressed upon the subject of the gold regions.

Divesting these accounts of certain expressions bordering rather much upon the hyperbolic order, they amount to the fact that the outcrops of certain veins have been removed. Such expressions might have materially increased "the fever," but for the frequency of similar causes which at length but slightly affect the body politic, because, like the body corporate in certain cases, it is becoming "acclimated." Some of the expressions alluded to, and copied from California papers into our own, about "*gold-bearing quartz said to be found in inexhaustible masses or quarries through the whole mountainous region which forms the western slope of the Sierra Nevada*," and "*these quartz-mountain quarries*," and divers others, are indicative of a state of aurimania. Accounts are also given of the yield of gold said to be averages of these great *gold* "*quarries*." That the specimens from which the gold was extracted contained the stated proportions is most likely, but that is a very different affair from the *average* rate of productiveness of a vein. The stories are altogether too flourishing; but making all allowances required in the premises, it is gratifying to me to find that my views in reference to the geological position of the gold, which were freely made known during my reconnoissance through the country, are now fully verified.

In trying to get at the substance of the facts through all this golden magnificence, it would appear that the outcrop has been removed from one or more veins which are likely to pay for the working, and *may* prove profitable if well managed. They do not appear to have penetrated beyond the outcrop, at least so far as to have determined the thickness of the veins, which can only be ascertained by penetrating into the hill until the slate or other rock *through which the veins* run is reached, and which, in the case of slates, is often covered with a considerable thickness of detritus. The cause of this, as well as that of the *quartz* veins, projecting usually *beyond* the rocks which they traverse, was explained in a former place in these pages.

It is very possible that outcrops of considerable thickness have been uncovered, as might be expected from what was also said in this memoir in reference to the quantities "of quartz seen stretching in lines over the country." These openings having developed thick veins, have given rise to the phrases just now quoted, and others equally extravagant, and would, no doubt, convey the idea to many that gold is expected to be quarried very much as we do building stone, and that this gold-bearing quartz constitutes great "mountain masses of rock." The application of the term "quartz-*rock*" to these *veins*, would not be made by one the least conversant with geology; in this sense simply it·is the *veinstone*.*

---

*Note.—Owing to causes not under my control, a delay has occurred in reporting the aforegoing memoir.

Since it was brought to a close we have accounts from California up to the first of January, some of which it is proper to notice.

The stories about "vast *formations*" of gold-bearing "*quartz rock,*" extending throughout the Sierra, are repeated with *ample* additions. We are told that "*great quarries*" have been opened in them, and that this quartz rock has been fully proven to yield a proportion of gold equal to from $2 50 to $3 in value to the pound.

The misuse of terms need not be *again* alluded to ; but I shall give such data as will enable any one, whose imagination does not outrun his judgment, to prove incontestably that these relations are altogether incorrect.

Referring to what has been already said in chapters I, IV, and V, in relation to the ravines and veins within the gold region, it may now be stated:

1. That the veins of quartz containing gold were filled before the' ravines were excavated, and that they extend over great distances.

2. This is evident from the fact of their outcrops being visible on the opposite sides of said ravines.

3. As the gold is nearly 20 times heavier than water, and 7 or 8 times heavier than quartz and other earthy matters of that region, it *must* remain in the ravines at little distance from the vein itself, except the *very small* proportion reduced to powder or minute scales that may have been carried down the streams, adhering to the earthy matters.

4. This is further proven by the fact that the most diligent searches of the diggers have failed to find it in quantities worth working, *below* or westward of the limits of the slate rocks in which the auriferous veins occur.

5. If the gold in any ravine were exhausted, and its exact value could be ascertained, we might readily determine, by measuring cross sections where it intersects the veins, as well as the aggregate thickness of those veins, the exact average proportion of gold contained in them.

6. The converse of this is equally certain. If we know the thickness of any vein, and the proportion of metal in the *veinstone*, it would be very easy to calculate the amount of gold that, because of its great specific gravity and indestructibility, would be under and among the drift matter.

7. Metallic veins vary so much in thickness and composition, that it is only after a vein shall have been worked for a great distance, or in numerous equi-distant places, to a considerable extent, that any one conversant with the subject could with propriety venture to state its average rate of productiveness.

The gold stories having at last assumed a tangible shape, it may be well to test them by the application of mathematical and physical science.

Where I crossed the north fork of the American river, its ravine is not less than 800 yards deep, and about 3,000 yards broad at the top, as shown in plate II, where *d d* shows the present line of outcrop of a vein which, we will suppose, runs at right-angles to the course of the river or the ravine.

We shall assume, for facilitating the calculation, the destroyed portion of the vein to be the triangle A, B, C, whose longest side (A B) is 3,000 yards, and the perpendicular from the centre of said line to C = 800 yards; therefore, $\dfrac{3000 \times 800}{2} = 1{,}200{,}000$ yards for the cross

section of the removed portion of the vein (*d d*) in the ravine. Consequently, if the vein be taken at only one yard thick, the removed portion which formerly extended (with the rocks it intersects) across the space now converted into the ravine must have amounted to 1,200,000 cubic yards.

Now, the specific gravity of the kind of quartz which constitutes these veins being 2 6, a cubic yard will weigh 4,320 pounds ; and, taking the lowest amount ($2 50) given in the recent accounts as the average product of gold in these veins of quartz to the pound, we have 4,320 pounds by $2 50 = $10,800 for the value of gold to the cubic yard of veinstone. This sum multiplied by 1,200,000, as above ascertained, gives $1,296,000,000 as the value of the gold formerly contained in the portion of the vein removed, whilst the ravine was being excavated to its present dimensions.

Now, during nearly two years these ravines have been diligently explored by gold seekers. Within the past year they have spread themselves along every ravine within the gold region from Feather river, and even further north to the Maripoosa, and other affluents of the San Joaquin, near its headwaters.

Within this metalliferous region of more than 1,000 square miles, there is not one square perch, perhaps, of drift in the ravines that has escaped the notice of the enterprising gold seekers. They have dug and delved everywhere to the bottom of the drift in search of the "rich diggins" they were induced to believe they would find. They have worked the harder from being

The object of thus referring to these *enlarged* accounts is to correct them, so far as can be done without revisiting the region, and to put the matter in such a light as may be realized by our countrymen disposed to emigrate to that part of the globe.

PHILIP T. TYSON.

FEBRUARY 20, 1850.

—

Explanation of the plates.

The map embraces all those parts of Upper California which I visited, except San Diego. The routes over which I travelled are indicated by dotted lines. The rocks and minerals seen are indicated by their names.

Plate I. Geological section from the Pacific ocean, across Bodega bay, to the summit of the Sierra Nevada, about N. 80° E.

This section, as well as the following one, is intended to aid in forming a general idea of the geological structure of a belt of country between the Sierra Nevada and the Pacific ocean, near the parallel of 38° N. latitude.

The firm lines upon the western edge of the Sacramento valley show the position *observed* of the sedimentary rocks. These, as stated in the text, are sandstones, indurated clay, and conglomerates, probably of the miocene period. The dotted lines indicate their position (as may be inferred) beneath the more recent deposites under the valley.

Plate II. Section from the Pacific ocean, in latitude 37° 30' N., in a course about N. 70° E., to the summit of the Sierra Nevada.

The firm lines here also indicate the position of the sedimentary rocks, which, upon this line of section, were observed upon *both* edges of the great valley, leaving no doubt whatever of the continuance of the strata under the valley, as shown by the dotted lines.

The industrial importance of this fact, in reference to supplies of water by means of artesian or other wells, was explained in the preceding pages.

Both of these sections cross the western slope of the Sierra Nevada, and, of course, the gold region.

The precise boundaries of the different formations crossed by these sections are not put down, because there was no time to determine them whilst merely making a reconnoissance through portions of the country.

---

impelled by absolute necessity to procure, if possible, at least a sufficient amount to support life. If these stories had been within the *bounds of possibility*, there would have been no other means of ascertaining whether they were correct or not, other than by a most careful systematic exploration performed on the ground.

Now, the outcrops of the auriferous veins, as before stated, are indicated by the lines of *angular* fragments of quartz. They are sometimes in close proximity to each other; but if we assume, as a mean, that they are a mile apart and only one yard thick, there would have been accessible to the gold digger in the numerous ravines of this region a quantity of this metal many thousand times greater than has been obtained from all sources since the days of Adam.

The aggregate *length* of the deeper ravines, where streams flow entirely across the gold region, is more than 2,000 miles; so that, with the data given, any one who desires to do so can calculate how much gold there ought to have been among the drift in the ravines of the Sierra, if the proportion of metal were as great as we are recently told. The relations about "great mountain quarries of gold-bearing quartz rock," and of its being "as abundant as anthracite coal in Pennsylvania," prove *too much* in admitting that the outcrops seen indicate veins of great thickness.

We have shown that the destroyed portion of a single vein, only one yard thick, ought to have left near at hand, among the drift, nearly thirteen hundred million dollars in gold; but, as they are called "great quarries," they have, perhaps, a thickness of many yards to swell the amount.

The positions of the principal rocks and minerals which were *seen*, are indicated by their names only, as on the map.

Plates III. IV. V. and VI. show the various rocks met with on lines of travel through the gold region.

Plate III. Section from Bear river, N. 20° E., about 20 miles, to the Yuba river, and continued south of and paralled to the Yuba, about N. 65° E., 20 miles.

Plate IV. Section from the Yuba, about S. 40° E., 40 miles, to the South Fork of the American river, at Coloma.

Plate V. Section from the Cosumes river, near Daler's Mill, about 25 miles, to the Calaveras river. General course, about E. S. E.; but the middle portions of the line are curved to the eastward.

Plate VI. Fig. 1. Section across a valley in the coast range, east of Livermore's ranch, illustrating the remarks in the preceding pages, upon the subject of obtaining water by means of artesian or other wells.

The sedimentary strata may be observed forming a synclinal axis at *a*, towards which the underground water flows. It·is· to be expected that water would flow above the surface in all such valleys, and supply ample means for irrigation.

Fig. 2. Section across a valley between Armidor's and Benicia. The strata under and on both sides of this valley dip uniformly to the eastward, and consequently there are no springs on the eastern side of the valley, or on the adjoining hills, (*b*;) whilst on the western side towards which the strata dip, there are numerous perennial springs, at various elevations, from the edge of the valley to at least 400 feet above it on the hill (*aa*) on the left. See page 65.

Plate VII. Diagram in illustration of the subject of surveying public lands containing valuable minerals. The veins (V) are intended to illustrate the position of a system of veins, if they were "stripped" of their outcrop and detritus. The *strings* projecting from veins are in some instances marked S.

The region from which this diagram is taken is near and between Jackson's and Sutter's creeks, as seen on the map. The topography is partially given, for the reason that there was no opportunity to sketch it on the ground; but enough is given for our present purpose.

The dotted lines indicate the mode in which the lands should be laid out, so that every lot would be *available*, by enabling the operator to push his work into the veins horizontally or upwards, so that the mines will drain themselves; and the mined matter may be taken out upon level or descending tram roads.

Upon the supposition that the veins are about one mile apart, the space embraced would be divided into eighteen lots, whose average content is equal to about 500 acres. The areas, of course, would be greater or less, according to the distance the productive veins may be from each other.

The firm lines show the effect of dividing the same area into lots of one mile square, or sections of 640 acres. This mode gives twelve lots within which the veins may be opened in the deep ravines, and, of course, advantageously; and six others whose veins could only be worked by *deep mining*, and pumping out the water—a serious item of expenditure everywhere, but which becomes excessive when the cost of fuel is considerable.

By dividing these lots into quarter sections of 160 acres each, it may be seen that about one half of them would be altogether useless, except to the owners of lots having fronts on the larger ravines, as explained on page 66.

Plate VIII. Section near Bodega Point, (enlarged,) in the line of section 1 : *a*, leptinite ; *b*, sedimentary beds, from which pure water flows in copious streams.

Figure IX. A section across the North Fork of the American river, referred to and explained in the note on page 71.

worke
expen
of fue
By
be se
to the
on pa
Pla
sectio
flows
Fig
refer

Pl. V.

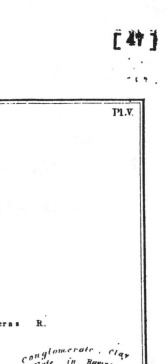

before the ... ion of the ... d, and the ... queror. The chief military ... esident of the United States, and in conformity with the laws of nations, governed the territory as a conquest. In the mean time a treaty with Mexico transferred the whole of the sovereignty and domain of Upper California to the United States, and guarantied to the people all civil rights conformable to our constitution.

[47]

74

work

expen

worl
expe
of fu

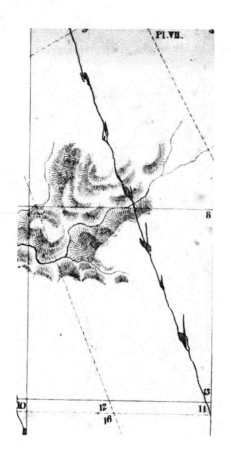

Pl. VII.

[47]

wor
exp(

Pl. IX

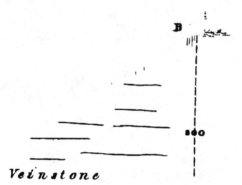

B

860

Veinstone

REPORT OF GENERAL PERSIFOR F. SMITH.

———

HEADQUARTERS PACIFIC DIVISION,
' *Fort Vancouver, October* 7, 1849.

SIR: I have the honor to address to you, for the information of the major general commanding in chief, and of the government, the following report:

I have from time to time previously reported events as they have occurred, with some observations on them, but shall now recapitulate them briefly, and enter somewhat at large into the peculiar state of affairs here, and explain fully the reasons which governed my actions in them—which affairs, though at first view not embraced in my duties as commanding this division, will be found to have much influence on the condition and service of the military force here. Besides, the law not having provided officers in other branches of service here, the government can only look for authentic information to those whom it has been authorized to send.

Under orders to take command of the "Pacific Division," I left Baltimore on the 24th November, and New Orleans on the 18th of December last, by the isthmus of Panama, for San Francisco.

While at Panama I learned that the worst part of the population from many of the Pacific ports was going to California to search for gold. As this could only be done by trespassing on the public lands, in violation of the laws and to the injury of the rights of the United States, I sent notice to those different ports that on my arrival in California the law would be enforced. The law itself points out the mode of its execution, which is through the courts of the United States. But as the last session of Congress passed without providing such tribunals, the law has had to remain inoperative, and must continue so till Congress provide otherwise

Some of the individuals offending have applied for protection from violence threatened by others, but I have informed them that I cannot interfere to secure them in the infraction of the law.

I arrived at Monterey on the 23d of February; and being detained a few days by the steamer's want of fuel, I had an opportunity of conferring with Colonel R. B. Mason, commanding the department.

He had been in continual expectation of hearing that provision for a territorial government had been made, and consequently had done nothing to alter the condition of affairs existing at the conclusion of the treaty.

In order to explain the object and operation of the measures pursued after my taking command of the division, I will state my views of existing circumstances, and their consequences at the time.

Upper and Lower California had been, for many months before the conclusion of the treaty was known, in the exclusive occupation of the forces of the United States. All military opposition had ceased, and the inhabitants submitted to the rule of the conqueror. The chief military commander, under instructions from the President of the United States, and in conformity with the laws of nations, governed the territory as a conquest. In the mean time a treaty with Mexico transferred the whole of the sovereignty and domain of Upper California to the United States, and guarantied to the people all civil rights conformable to our constitution.

Congress having been in session at the conclusion of the treaty, it was supposed they would provide a territorial government for California; in expectation of which, no steps were taken to change the existing state of things then, though it required attention; for, during the revolutionary movements which preceded the occupation of California, many of the higher civil offices of the department were vacated, and the regular administration of the law interrupted; and while the military government under our conquest existed, many of the inferior Mexican officers of justice were displaced as inimical. So that the termination of the session of Congress found and left the Territory almost without law. And although this might have endured without becoming intolerable previously, when the inhabitants and interests were few; yet now, the discovery of the gold mines had already given a strong impulse to commerce and emigration, and transactions involving much property and many rights were calling for the aid and protection of law, when the news arrived that Congress had adjourned without any legislation on the subject.

The people of the Territory finding no relief from Congress, and understanding from an expression in the message of the President, urging the passage of a territorial law, that, unless Congress acted, the people had no law—*de jure* as well as *de facto*—thought their only remedy was to create a new government; ignorant, as most of them were new comers, that sufficient and excellent law existed, if it were only revived and put in motion by the appointment of proper officers, whose absence they then took for the absence of law. They proposed following the example of Oregon, to create a provisional government; and some districts had chosen delegates, and a time and place were fixed to hold the convention. Some, too, justifying themselves by the natural right of "self-government inherent in man," had ordained themselves into separate communities, established codes of laws, and appointed officers to execute them; disposing of public and corporation lands, and claiming the revenue collected from commerce. Though I had no authority to create or organize a government, or dictate how or when that should be done, yet, as my instructions and duty required me to aid the civil authority, when called on, in support of the law, it was absolutely necessary to ascertain what was the law, and who were the legally constituted authorities in the Territory.

Having satisfied my own mind on these points, I urged the same reasoning that influenced me on all those with whom I had intercourse; and events have since taken a direction in conformity with it. I will state it as concisely as possible:

First. In relation to an existing government in California.

Previous to the Mexican war, Upper California, as a department of the republic of Mexico, had a completely organized government, (in theory) as complete as that of any State in our Union, though, in fact, irregular and disordered in its action from internal commotions. It possessed an executive, a legislative body, a judiciary, subordinate executive officers, and all the machinery for administrative and financial operations; connected with the general government through the military officers of the latter, and by the right of appeal and supervision.

When this territory was conquered by the United States, the latter acquired the right of changing and administering this government so as best to conduce to the objects of the war.

By the treaty we made California a part of our Union; and thus, by another title, acquired the right to regulate, as provided in our constitution, its government.

But neither the laws of nations nor reason require, or permit, that an acquisition by conquest, much less one by treaty, should, *ipso facto*, abrogate the civil institutions of the country acquired. Those relating to allegiance and sovereignty are of couse modified, for these are the very objects of the transfer; the others remain until altered by the new sovereign, in the way pointed out in its own institutions.

California, then, since the conclusion of the treaty, and now, is governed by the laws and institutions she at first enjoyed; modified, as they may have been, during the conquest, by the conqueror, and as they have been since the annexation to us, by the effect of the laws, constitution, and treaties of the United States, of which she now forms a part. And to give her the practical benefit of these laws and institutions, it is only necessary to put them again into motion. Some, indeed, that require the concurrent action of their general government must remain inert, because there is no law of the United States to point out how they are to be exercised. These laws, though still operative in California, cannot affect or control the Executive or Congress of the United States.

Secondly. How can these laws and institutions be changed?

It was urged that, granting the old laws of California to remain, yet the people of this Territory, if dissatisfied with them, can change them; that the fundamental right of self-government, if it allows a community to be formed, and to establish its own laws, allows it also to change them whenever they become dissatisfied; consequently, the people of California, if dissatisfied with theirs, as different from those under which they were bred, may, by general consent, alter them; and they cite the example of Oregon.

To this it was answered, that Oregon had no previous government; there were no laws in existence there.

The action of our own government in the conventions with Great Britain for joint trade there, seemed to throw a doubt upon her own right to establish any government; and, if the people of Oregon could claim the protection of no government, they owed allegiance to none, and were therefore free to listen to their own necessities, and create one for their own protection. They had no laws, for they formed part of no State; a Congress had passed none for them, so that the first want of society justified their action. Granting, even, that the reasoning was sound in relation to Oregon, it could not apply to California, for her circumstances were entirely different. She had, and has yet, a complete system of law, founded on the best in the world, the old civil law of Spain. She is part of the United States, under the protection and influence of its laws. And this fact determines in whom the right of legislation lies.

No citizen of the United States questions the original right of a yet unorganized community, owing no allegiance elsewhere, to establish and regulate its own government; and if the lands, sovereignty, and allegiance of California have no claimants, the present occupants could for their own safety and comfort do it. But the people of the United States, as communities, have already exercised that right; they *have* established governments and laws. Those laws for civil and personal rights they have generally reserved to their own control in the several States; those of general and exterior interest they have committed exclusively to an agent, in the shape of a general government. Every case is provided for in the States respectively; each regulates its own affairs. In matters of general interest, and in the common property of all, they have committed all legislation to Congress; and through one or the other of these means alone, can any acts of government or legislation be done within the limits of the territory of the United States. And so in Oregon; the validity of their provisional government depends on this fact—was it or not a part of the United States? If it was, their only remedy is to claim of their government redress for the evils inflicted by its neglect; if this neglect cannot be justified, they cannot usurp the rights of the people of the United States. But the people of California have not even this doubtful point to rest on. By solemn treaty California *is* a part of the United States, and its people have no more right to this independent action, than have the inmates of an asylum belonging to the city of New York to erect themselves into an independent community or corporation and dispose of the houses and land they occupy.

But these same fundamental laws of the United States permit the people of California, with the consent of Congress, to form a State in the confederacy, and thus acquire all the rights of sovereignty not bestowed by the constitution or Congress.

If they do not elect this alternative, the other alone remains open to them, to wit, to await with patience the action of Congress; for, if it possesses alone the power, it does also the discretion as to the mode and time of legislation, and may postpone the interests of the people of this section to what may be considered the more important interests of the whole.

Some urged that the formation of a "territorial government" was a necessary preliminary to the existence of a State. To this it was answered, that such a thing as a territorial government is unknown to the constitution. It is merely the mode by which Congress, for convenience in certain cases, have chosen to exercise the duty of "legislating for the Territories of the United States."

In accordance with these views, I declined, when called on, to recognise any authority not existing by the laws of California or those of the United States.

At the time of my arrival at San Francisco, (February 28,) however, the session of Congress was again nearly ended, with every expectation that this subject had been acted on. So that, to avoid the trouble and expense of re-establishing the former authorities, to be replaced probably before they were fairly in operation, I thought it better to wait, and directed the steamer Edith to be sent to Mazatlan and San Blas, from which, by the Vera Cruz mail, the earliest intelligence of the action of Congress would be received. Upon her return with the news, Brevet Brigadier General Riley, who had relieved Brevet Brigadier General Mason, proceeded to revise the administration of law under the previously existing institutions, and recommended the people to choose delegates to a convention to form a State, to meet at Monterey the 1st of September.

I will here mention that, it being impossible for private persons elected as members to find the means of conveyance from the distant parts in the present state of the country, and as their mileage would probably come ultimately from the national treasury, I authorized the quartermaster to provide transportation by water for the delegates distant from Monterey.

In adopting the course I have pursued in this subject, I have been governed by what seemed to me the importance of arresting the construction of a system which, however excellent it might be in itself, had the inherent and fatal vice of nullity at its foundation. The laws passed under it in some places were undoubtedly good, and would have been honestly administered; but as large commercial interests, not only with the merchants of our Atlantic States, but with all the world, were creating, and must be subject to the operations of these laws and of the courts established under them; as real and personal property were to be sold, successions administered, and judgments, civil and criminal, to be executed by their order, involving much money and the interest of government, and many absent persons, and possibly the lives of some here, I thought it of great importance that nothing should be done the validity of which was not only doubtful now, but the defects, when ascertained hereafter, incurable, and which must throw all titles and contracts into perpetual uncertainty.

Nevertheless, no military authority or force has ever been exerted in opposition to any person or thing not subject to military law. Nor, to my knowledge, has there ever been any the slightest dispute or collision between the army, or any part of it, and the citizens; on the contrary, the greatest harmony has prevailed. And I doubt whether any part of the United States has presented a community in which there has been so few crimes or even disorders committed.

The public records of any of our largest cities will present more of these in any one day, than have taken place in the whole of California since my arrival.

A riot in San Francisco, which the citizens themselves suppressed, and an outrage on Indians in the unsettled country above Suter's fort, by strangers to California, are the only ones I know of.

Many misrepresentations have been published on this point. But the people of the United States may rest assured, if California ask admission into their confederacy, that no community is more able and desirous of maintaining peace and order.

This concludes what I have to say on the civil organization of the Territory, as connected with my duties.

Another point on which there was embarrassment from the want of legislation by Congress was the collection of the revenue on imports.

This was a matter exclusively governed by the laws of the United States.

Presuming that Congress, after the conclusion of the treaty with Mexico—which had incorporated California with us—had provided for executing the revenue laws here, Colonel Mason directed the collection of duties according to those laws. On my arrival here, knowing that no such provision had been made, I directed a different course to be pursued.

The circular of the Secretary of the Treasury had declared that, by the treaty, California was part of the United States, subject to all its general laws, and vessels could trade coastwise to and from it with other parts of the United States. But no provision was made for the appointment of collectors, or designating ports of entry.

Nevertheless, an immense amount of shipping was on its way here.

The revenue laws provide that all goods of a certain description brought into the United States shall pay certain duties. Consequently, they cannot be brought here without paying such duties. But these duties cannot

be paid, because there are no persons here provided by law to receive
them; so the goods cannot be introduced. Vessels with such goods can
only go where there are collectors to receive the duties, enter their goods
and then send them coastwise here in American bottoms. The nearer
collector, then, was in an Atlantic port; and as all crews deserted on their
arrival here, vessels once arrived could not get away. And no vessel was
provisioned for a voyage to the Atlantic. Besides, the people here were
in distress for many articles brought, for the country was bare of goods.

To meet these circumstances, I directed that the consignees of dutiable
goods, as duties could not be exacted of them, should have the option to
comply strictly with law and send away their goods, or deposite with the
person acting as collector the amount that would be due under the revenue
laws, subject to the action and decision of Congress. This has hitherto
been done; and the amount thus paid has been placed in the hands
an officer of the Quartermaster's department, as depositary, and its e
position will require the action of Congress. As the accumulation of a large
amount thus not only embarrassed commerce, but incurred much risk,
the disbursing officers were authorized to use it for purposes provided by
law, in place of drawing money from the United States; thus distributing
it for circulation, and, in reality. transferring it to the treasury for safe
keeping until Congress should dispose of it. In the mean time, lest the
conditional admission of goods should be misunderstood as a regular
opening of the ports, I addressed a circular to our consuls on the Pacific,
advising them of its nature, and that it might at any time be disapproved
of by the government, and that shippers must send goods here at their
own risk of having them refused entry. This was done that government
might revoke the privilege granted without being annoyed by claims for
indemnity. But a law had passed providing collectors, though the officer
appointed for San Francisco had not arrived when I left.

The accounts of the collectors are returned to the commander of the
department, and by him, I presume, transmitted to Washington. I have
no return of the amount collected, and which requires the action of Congress.

In relation both to the organization of the civil government and the
question of duties on imports, much stress was laid by some on the
repugnance of the people to "military government."

To the first it was answered by the laws of California itself—not by
any usurpation or imposition. Where there is no governor to the department, his functions are to be exercised by the military commandant of
the department.

This designation by law of a military officer to perform civil functions
no more makes such functions military than would his being drawn to
sit on a jury in a civil court make it a court martial. It is the princi-
ples and rules of action that give its character to the government, and
not the other attributes of the persons who administer it. As to the revenue
laws, there being no other than military officers here, they were bound
to see that the laws of the United States were not infringed, and to secure
the rights of the government under its laws until it made all dispositions
necessary; and what was done was extending favor to the people here,
not restraining their rights.

These distinctions have been most carefully preserved, and in no instance in my knowledge has any of the executive power been authorized

to act except through the civil branches of the administration. I have been careful to urge this distinction in all things—that those duties under the laws and authority, or affecting the rights and interests of the United States, should be exercised through the regular military staff of the department, while those connected with the civil administration of California, under its own laws, should pass exclusively through the hands of the civil officers and the forms of the civil law.

It will be seen, from what I have above stated, that the only revenue collected must accrue to the United States, for there has been no departmental legislation to lay taxes, and there is no fund to pay the necessary expenses of government except what may have remained in the treasury of the department at the conclusion of the treaty.

A population of at least fifty thousand people—a yearly product of near ten millions of dollars—a commerce employing several hundred ships, their cargoes mostly belonging to merchants in the Atlantic States—the most active trade—the security of so much property, and the lives of so many people,—are such powerful reasons for establishing a satisfactory government that can make all these resources available, that I presume, that if it becomes necessary to apply any portion of the money not properly belonging to the department to this end, Congress will approve of its proper and economical application by making the necessary appropriations.

At the time of my arrival in California, there were but few troops remaining in some of the posts occupied during the war.

The troops in that department now consist of the second infantry, two companies of the third artillery, and three companies of the first dragoons.

In selecting positions for them I at first fixed upon St. Luis Rey, an abandoned mission in the southern part of the Territory, as the location for the greater part of those expected to arrive. Colonel Mason represented it as in a district fertile though not extensive, and central as to San Diego, and those points on the frontiers by which our boundary with Mexico was crossed, and that the buildings, with a few repairs, were sufficient for the troops.

It appears, however, from subsequent information, that the cultivation has been neglected and the neighboring inhabitants gone to the mines; that stragglers had taken possession of the buildings and rendered them uninhabitable, so that other dispositions were made.

The posts now occupied are San Diego; Monterey, the headquarters of the department; Presidio of San Francisco, two miles and a half north of the town; Benicia, the general depôt; Sonoma, the headquarters of the division; a post on the Stanislaus river, twenty-eight miles southeast from Stockton; a post on Bear creek, ten miles above its junction with Feather river, about thirty miles northeast from Sutter's fort. The command destined for this post, under Major Kingsbury, had reached the neighborhood of Sutter's fort in June. It was still there when I left California, the first of September.

General Riley proposes to have the post on Stanislaus river removed further south to King's river. According to my present information, a post there will be very difficult to supply; but a careful reconnoissance, without which no change will be made, will determine that point.

The troops on Bear creek and Stanislaus river may possibly be withdrawn to the coast during the rainy season, for the country immediately

below them will be then impassable for wagons, and will be overflowed
when the mountain snows melt in the spring; they may likewise prove
unhealthy at this season, in which case they will be moved further into
the mountain—they are now just on its last declivity towards the plains.
These considerations, and the difficulty and expense of erecting barracks,
will suspend the permanent location of these posts until more experience
of the climate and seasons is procured.

The object of these posts is not to maintain garrisons large enough to
make any important operations in the Indian country beyond them, but
rather to serve as advanced depots for supplies for corps that may move
in that direction.   Above them the mountains may be travelled at all seasons as far as the snow, with portable india-rubber boats to cross the
streams: below, between the coast range and the mountains, in the valley
of the San Joaquin and the Sacramento, the country is impassable for
animals in the rainy season; and if troops are to move, then they can go
in boats through the low countries, and find their pack-animals and supplies at these depots, which will have been furnished in the dry season.

The post at Sonoma is occupied by Captain A. J. Smith's company,
1st dragoons: this company is too small (less than a dozen men) to occupy any distant post. I have used it as an escort in my journeys through
the territory. When I left Sonoma, it was under orders to visit the country north of Sonoma, where some Indians were troublesome, and to explore the country beyond. Sonoma is at the outlet of one of the valleys
leading from San Francisco bay northward, between the coast and the
Sacramento valley. When more troops are disposable, the post will be
advanced more to the northward.

The general depot for the division is at Benicia, on the straits of Kar-
quines, at the junction with Suisun bay, and on the north point of the
straits, fronting on them about half a mile, and about twice as much on
the bay.

The town of San Francisco was noways proper for a military depot;
there is no harbor there more than in any part of the bay, which is too
large, and consequently too rough for the loading and unloading of vessels. The landing is difficult, and can only be used at certain time of
tide. Its position is unsafe, lying with its back to the seacoast, on a
narrow peninsula, cut off from the main land, except by a circuit of
forty miles. The lease of the stores occupied there was to expire in June
or July, and could not be renewed but at rates incredibly large; and the
expense of landing and shipping goods was so exorbitant, that economy
was consulted by quitting the place altogether and establishing it elsewhere.

I selected the present location because the harbor is safe; being in fact
a river five miles long and one broad. Vessels can unload by a stage from
their side to the shore. It has a good communication by land and water
with all parts of the country: the largest ship of war can run in one tide
from sea to this point, and it can be perfectly defended. The persons
claiming the land under a Mexican grant have ceded their property in it
to the United States, provided it be used for the purposes designed.

I have directed storehouses for the quartermaster's and commissary's
supplies, barracks for two companies, and quarters for the necessary officers, to be constructed in the cheapest manner, of wood. As the lumber
was to be got from Oregon, and labor was hard to be procured, I author-

ized a vessel to be bought as a storeship, the cabin of which might serve for officers' quarters. Enough of the buildings projected to secure everything from the rains of this winter are now, I presume, finished.

As the state of things in California renders it impossible to carry on any work except at the most extravagant expense, I have limited all operations to those of the strictest necessity.

The creation of a proper depot was most pressing, for we were subject to extravagant rents; and as our supplies would come but annually, a large quantity must sometimes be on hand, and, if injured from exposure, could not be replaced under a long time.

Next was some defence to the entrance of the bay of San Francisco. For this, I directed some guns lying on the beach at the town of San Francisco to be removed, and six or eight of them mounted on an old work on the southern point of the straits at the entrance of the bay. The repairs necessary were limited to the construction of a temporary magazine, constructing epaulment to cover two large guns in the gorge, and replacing the platforms. Major Ogden, of the engineer corps, here on other duty, furnished the plans and details of the work, and the execution of it was committed to the officer commanding the neighboring post. This work, imperfect as it will be, is of importance when there are no vessels of war here, for with the ebb tide and usual winds any vessel can escape to sea without risk of being overtaken.

The other point demanding immediate attention was opening a sure and easy communication by land with the Atlantic States. The whole value of these countries to the United States depends on this. The route by the isthmus is too expensive and too insufficient for the number of travellers. The steamers can bring with propriety not three hundred a month, while the emigration by land, if divided throughout the year, would average three thousand a month. The route by sea, either across the isthmus or round Cape Horn, uncertain and insufficient at all times, in time of war, when most needed, would probably be entirely interrupted. A route across the interior is practicable, because it is annually travelled. But the way may be made better and more sure by careful explorations. As these can only be made in the summer and fall in the mountains, and Congress might be disposed to act at once, I determined to employ what remained of the proper season this year in having examined that one of the various traces within the limits of my command that seemed to offer most advantages.

As the principal obstacle at this end of the road is the Sierra Nevada, the examination was confined for the present to the passage of that ridge from the low plains on this side to the high ones on the other.

On information and advice furnished by Colonel Frémont, through whom chiefly the interior of this country is known to the people of the United States, I selected for examination a pass by the Cow creek, one of the headwaters of the Sacramento. Brevet Captain Warner, an intelligent and accurate officer of the Topographical Engineers, was charged with the duty, and an escort under the command of Brevet Lieutenant Colonel Casey, 2d infantry, accompanied him.

They left Sacramento City in August, later than I intended; but the depot on Bear creek was not established, and the expedition had to be fitted out from resources at a distance. To avoid the necessity of a second examination, Captain Warner was ordered to provide himself with the instruments necessary to establish the practicability of a railroad by this or any adjacent pass, and has accordingly gone so prepared.

These three objects of primary importance were intended to limit the expenditures out of the usual military routine; but as I was about starting for Oregon, the emigrants from the United States began to arrive. Hitherto a few hundred had crossed annually, and had often suffered from the want of pasture for their animals, and all the consequent distress—for the loss of animals involved the abandonment of wagons, and loss of provisions and clothing. An early fall of snow not long since arrested the caravan of that year before their arrival near succor, and some of the survivors were found who had been reduced to the horrible necessity of feeding on the dead bodies of their companions.

As the road this year was crowded with from five to seven thousand wagons, it was probable that the suffering would be beyond description among those who, coming last, would find the whole of the latter part of the route bare of forage, and in many places everything burnt off by fire.

Some of those who had already arrived represented the situation of many whom they had passed as destitute in the extreme, and begged that food and assistance should be sent to them. Under these circumstances, I directed as many wagons as could be procured up the Sacramento to be despatched partly loaded with provisions, and to proceed in two parties by the two routes on which the emigrants were advancing; to issue provisions to such as were in need, and to send in the women, children, and sick in the wagons emptied. No military escort was considered necessary, as the road would be full of emigrants, and no troops were disposable to compose it. Brevet Major Rucker, 1st dragoons, was selected to take charge of the train, and carry out these dispositions. I left California immediately before any of the preparations were completed. The necessity of the expenditure in this matter, if any was incurred which is not authorized by law, will, I hope, be a sufficient reason for making the requisite appropriation.

The troops in Oregon consist of eight companies of the regiment of mounted riflemen, under Brevet Colonel Loring, commanding the department, who has just arrived, and two companies of the 1st artillery, which arrived, under the command of Brevet Major Hathaway, in May. Major Hathaway's company, L, is stationed at Fort Vancouver, on the Columbia, a trading post of the Hudson's Bay Company. Company M, Captain B. H. Hill, is posted at Steilacoom, on Puget's sound, eight or nine miles southwest from Nesqually. Two companies of the rifles, under Brevet Lieutenant Colonel Porter, were left at Fort Hall. Brevet Colonel Loring, with the six remaining companies, is now in quarters in Oregon City, at the falls of the Willamette—the only place in the Territory where houses could be procured to cover them. In compliance with instructions I had previously sent, a party, under Lieutenant Hawkins, of the rifles, who had arrived in the spring with Governor Lane, in command of his escort, was sent to Fort Hall, with provisions, to meet the rifles. They, however, took different routes, and did not meet. If he arrives at Fort Hall, Colonel Porter will have provisions for the winter. If by the 1st of November provisions have not arrived, I have sent orders to Colonel Porter, through Colonel Loring, to fall down to the Dalles, on the Columbia river, and occupy the abandoned mission buildings there, at which point the troops can be supplied from Fort Vancouver, most of the winter, by water.

If a post were established at Fort Hall to assist emigrants, it would be nearly useless, because they now follow a new route more to the southward. The country produces nothing but grass in the summer. The nights are so cold, from its altitude above the sea, and proximity to the mountains, that nothing that serves for food will grow there. The difficulty of reaching it from this coast is much greater than from the other, though the distance is less. The road to Oregon is made to *come* here, not to return; the ascents were chosen with care to bring up the loaded wagon of the emigrant. On the descents this way, they took the shortest lines; and though a wagon could slide down coming this way, which was all that was wanted, it could not be pulled up going the other. There are no Indians in that neighborhood particularly requiring restraint; and, if there be no particular reason unknown to me why a garrison should be kept there, I would urge that it be withdrawn from that neighborhood until the country is examined and roads opened, by which posts can be supplied. No doubt as good roads can be found going east as coming west.

My intention was to occupy but few posts at first; but every summer, while the pasture was good, to send strong detachments with pack-horses to examine the country and select routes for roads, and, when our knowledge of the interior should be more complete, to extend the posts as may be necessary. The influence thus to be exercised over the Indians will be much greater than by subdividing the force in the Territory, already small, into many and widely scattered garrisons.

The information given by travellers, or even by hunters and trappers, is not to be depended on; they only search for or observe what interests their own concerns. The contradictory nature of their reports, and the evident manner in which each makes his story conform to his own plans, is the best evidence of the little dependance to be placed on them. Travellers, especially, see nothing but the track they travel on: if it be sufficient for their purpose, they seek no other. A careful reconnoissance, by an intelligent officer with a party strong enough to secure his traversing the whole country, is the only reliable source of general and correct knowledge of topography.

I had intended to establish a post on the Upper Willamette river; but the communication from it to California would be so easy that, under present circumstances, it would be unadvisable. In the spring, when the rains are over and the men can work in the woods, the company of artillery now at Fort Vancouver will go to Astoria, and put up quarters for itself. Two companies of rifles will be stationed at the Dalles, and the remainder will leave Oregon City and take post at or near Vancouver. This is the only place in the Territory where there are buildings suitable for storehouses, and it is well situated for the principal depot of the department.

All the posts in the two departments are accessible by water, except Fort Hall. The whole country at the same time is very mountainous. The difficulty of making roads by land, and the fact that, when made, many will be impassable during the latter part of the rainy season—that is, from Christmas to the 21st of March—point out the necessity of providing the surest and most economical means of water transportation.

That part of the Territory included in the Pacific division, most likely to be occupied and settled within the next twenty years, extends about a thousand miles along the coast from San Diego to the straits of Juan de Fuca, while it is not two hundred miles broad from the eastern slope of the Cascade mountains and Sierra Nevada to the ocean. Puget's sound, the Columbia river, and San Francisco bay, afford access to the greater part of this Territory; but it is evident that the sea is to be the great highway between the different parts. Now, a peculiarity exists here to which we are not subject on the Atlantic coast. The winds blow continually from the northward, with rare exceptions; and a current, conforming with the direction of the wind, generally sets to the southward, so that passages from the south northwardly are very tedious, and in flat vessels at times impossible.

The coasts are rocky, the entrances to the harbors narrow, and fogs very prevalent at some seasons, for weeks together, so that sailing-vessels dare not approach the coast; because, if they should not make their port accurately, they could neither anchor nor haul off, but would drift on the rocks.

I was a month coming from San Francisco to Fort Vancouver; a steamer would have brought me in four days or less: yet we made a much shorter passage than any vessel immediately before or after us. And since my first arrival in the country, vessels have been twenty days coming from Monterey to San Francisco—about ninety miles.

This uncertainty and loss of time is more expensive than the employment of more costly vessels, in which these are avoided, especially when wages are very high.

The quartermaster's department should have two good sea steamers to run on the coast from Puget's sound to San Diego, and smaller river boats—one for the Sacramento and San Joaquin, and one for the Columbia river. The sea steamers should be very strong, both in hull and engine, because they will be exposed to heavy seas on the bars; and the engines of all should be powerful, to stem the strong current with which the tide runs.

There is plenty of wood for their use on the Upper San Joaquin and Sacramento, and on the Columbia river; and coal from Vancouver's island will be delivered by the Hudson's Bay Company at fourteen dollars a ton.

Without such vessels, the communication between the different parts of this division is too uncertain to be depended on.

All supplies sent from the Atlantic ports by sea should leave there by the 20th of November, so as to pass Cape Horn at the best season, and arrive on the coast in the most favorable weather of the year for entering the harbors.

From the middle of April to the 1st of August the entrance of the Columbia is most easy, for the weather is not foggy long at a time, and the sea is smooth. The rest of the year, a delay even of weeks must be provided for.

Steam-vessels are not embarrassed by calms or head winds, and in a fog can enter by the lead with more certainty than a sailing-vessel, as their rate of going is more uniform and direction more sure; they will meet, therefore, with much less delay at the mouth of this river.

I have been directed to select a position on the Columbia river for an ordnance depot. As the state of things in this country, to which I will presently refer, precludes the probability of getting labor for any but the most temporary public works, I shall merely select and have marked out and reserved the position having most advantages, without expending any money or labor on it. If the property of the Hudson's Bay Company should be purchased for the government, that at Vancouver will furnish all the buildings necessary for storehouses and depots of all kinds for many years to come.

I shall direct to be marked out and reserved positions for fortifications at the mouth of the Columbia and on Puget's sound, where there are no titles recognised by the treaty of Washington in conflict.

Oregon and California lie between the Rocky mountains and the Pacific ocean, from latitude 49° north to the river Gila—the 42d parallel dividing the two Territories.

From the Rocky mountains to the Sierra Nevada and Cascade range is represented to be a high, barren, rolling country, intersected with mountains, very bare of wood, generally of useful vegetation; in some places, especially of Oregon, furnishing abundance of grass for pasture, but generally producing nothing but wild sage, the temperature at night being too low for most articles of vegetable food; often destitute of water; here and there, but far between, some fertile valleys of small extent. This part of the country I have not seen, and only derive my knowledge from sources open to all. But it is clear, from universal report, that a large portion of this country offers no inducement to settlers; that there are some fertile districts of small extent; that there is pasture for immense herds of cattle.

From the eastern slope, Sierra Nevada, and Cascade range—to which district my observations hereafter principally refer—with the exception of the highest ridges of the mountains and the southern sandy part of the territory, the soil in itself is generally good; it possesses the elements of fertility, but these are only developed in proportion to the supply of water for natural or artificial irrigation. The rains begin in October; after Christmas are more plentiful, and end in March. From that time until October again, not a drop of water falls on the surface in California, and only occasional showers until July in Oregon.

In May the whole surface of the earth is covered with vegetation of the most luxuriant kind; grasses of various species, mixed with a profusion of flowers, as delightful for their perfume as for their beauty; but in a few weeks these are parched up, and die for want of moisture, and in midsummer the entire country out of the forests (and they are rare in California) looks like an arid desert of brown sand.

Those plants which have the advantage of the winter rains, and come to perfection before the moisture is evaporated, in summer thrive and are productive. So, our winter grains, potatoes, melons, and vegetables of quick growth, grow to great perfection; and where a stream of water affords the means of irrigation, the soil is productive during the whole summer. But the greater part of California is destitute of running streams and springs in the dry season. In the mountains the beds of the larger streams run in very deep, narrow ravines, whose sides are steep and rocky. Smaller streams near their sources in the mountains often water very beautiful and fertile valleys of small extent, but they soon sink into the soil and leave their beds dry. In the ranges near the coast, water is still more rare in California, though plentiful in Oregon; but the strip of land between the coast range and the sea derives sufficient moisture from the heavy fogs to supply all vegetation that is not injured by salt air in that part of it level enough for cultivation.

Generally speaking, California is bare of timber. The Sierra Nevada in its higher parts is covered with noble forests of pine, fir, spruce, and hemlock, with some oaks. Some of the coast ridges have remains of forests of "redwood," a tree partaking of the qualities and appearance of our cedar and cypress, and growing to an immense size. Some of the lower hills and plains have scattered oaks, resembling our live oak, but this timber is of bad quality.

Going northward from the bay of San Francisco, wood becomes gradually more abundant until you approach Oregon, where begin forests of pine, fir, and spruce, which extend to and beyond the Columbia, unsurpassed in size and density. So much of both Territories is mountainous to the very coast, and so much is without water, that not more than a fifth of the whole can be cultivated unless a supply for irrigation can be procured from wells. It is very desirable that machinery for boring artesian wells should be sent out, and employed in positions very desirable for military posts if water can be secured.

In stating the disadvantages that California labors under, it must on the other hand be considered that she has scarcely any winter, and her soil is not so unprofitable in her dry season as that of the Atlantic States is in their winter; and that the cost of irrigation where the water can be procured at all, is less than that of draining in Louisiana, since irrigating canals are on the surface, while draining ditches require to be cut deep below it.

In Oregon the want of water is not felt; but though the soil is generally fertile, much of it is encumbered with immense forests. Immediately on the seacoast, from cape Conception to latitude 49°, cold fogs and winds prevail in the summer afternoons; in San Francisco they are almost intolerable, while a few miles inland these are unfelt. With this exception, the climate of western California and Oregon is unsurpassed, and indeed unequalled, by any in the United States. The temperature is pleasant and regular, the nights always cool, no bad weather during the whole summer, and the winter so mild as seldom to bring snow in California, or to require in Oregon the cattle to be housed.

The largest potatoes, turnips, onions, beets, and radishes I have ever seen, grow on the Columbia river; and the best grapes in the United States grow in profusion in southern California, where some excellent wine is made.

Agriculture in these parts of both Territories affords inducements to the farmer, both of interest and comfort, unsurpassed, whenever affairs are so regulated that labor can be secured.

The lumber trade and fisheries in Oregon will one day be great sources of wealth, as will the former some future day in the mountains of California.

There is limestone in California in small quantities. I have seen it at New Almaden, (Forbes's quicksilver mine,) near San José, and have heard of it elsewhere. None has been found in Oregon, though it abounds in the now British territory north of 49°.

No coal has been yet found in either Territory; the specimens hitherto found are lignite, unless some lately discovered on the Yacona river, by Lieutenant Talbot, 1st artillery, should prove to be coal. I think it is not, but am not mineralogist enough to decide. He only found a small vein of it.

I transmit a copy of his report of a very interesting journey made to some of the small rivers south of the Columbia, conducted very satisfactorily and much to his credit.

There are large springs of mineral tar on the southern coast, especially near Santa Barbara.

There are many indications of iron and copper, but no ore has been found, worth working.

The vein of cinnabar at New Almaden is very rich, and apparently of great extent. It is only here that a vein has actually been opened. With regard to other places in the neighborhood where it is said to exist, nothing certain is known, as no sufficient examination has been made.

I have at different times visited that part of California north of Monterey, between the highest ridge of the Sierra and the ocean, except the district towards the boundary, and will have examined the country on the Columbia and Willamette, and across to Puget's sound. The southern part of California, and the country east of the mountain, I have not seen.

The month of July was spent in a journey through the gold region in the mountains, from the waters of the Yuba to those of the Towalumnes.

The discovery of gold mines in California is not of more importance, from their value and extent, than from the influence it has had on all public and private interests in the Territory.

The strip of country forming the greater part of the western slope of the Sierra Nevada, from the base of the central rocky ridge down to where it intersects the plain of the Sacramento and San Joaquin rivers, and from the headwaters of the former in the north, south to the sources of the latter, has been searched by adventurers with so much success, that it may be assumed that the whole district contains gold.

The gradual or convulsive changes, operated by nature in the course of many ages, have, in some places, protruded through the original deposite masses of non-auriferous rocks and earth; and in others, heaped on it impenetrable piles of volcanic or alluvial substance; but these, though extensive, are but partial exceptions to the general character of the country. And the lately acquired experience of the gold hunter has already taught him to avoid these, and to seek for the best reward of his labor in the neighborhood of the quartz and talcose slate ridges, or in the alluvion deposites drawn from them.

These last mentioned strata are found generally east of a line drawn through the junction of the forks of the American river, parallel with the mountain. The silent operations of decay have been for many years releasing the small particles of gold near the surface of these rocks, and the annual rains have carried them, in the earth in which they laid, down the precipitous sides of the hills to the streams; here the rapid current has carried along the earth, while the gold has settled down into the cavities of the rocky bed. Some of these streams have been filled up by slides of earth from the hill-sides; some by the deposites left in the still water formed by accidental obstructions; and the water has abandoned its old bed. Others still maintain their original channels. But the gold digger has learned to trace out even the closed water-courses, and, getting down to the original bed, finds, in what he calls the "pockets," the accumulation of previous times.

In some places, waters of considerable volume have been by the adventurers turned from their existing channels, and large quantities of gold found in the hollows thus exposed.

Enough gold has hitherto been found by thus searching near the surface to satisfy all those who labored with assiduity, and there has been no need to try excavations demanding much capital and time; so that the metal yet remaining in its original matrix is yet unknown. As there is no great certainty as to the amount already collected, there can be still less as to that which remains.

In the first place, the limits of the "gold district" which I have given above are only those within which gold *has been found.* Further examination of the mountain eastwardly, and of the country north and south, may extend there. Gold has undoubtedly been found in the coast range, north of latitude 39°. Next, but a small part of the gold district has been actually examined. And lastly, even when it has been worked, but a small part of the surface, and to the depth of but a few feet, has been searched. To give merely an opinion on this important topic, is to add nothing really to what is known. But I may venture to say that, considering the experience in other countries and the facts that are known here, I think that three or four years will elapse before the extent of the gold-bearing region is accurately known; that it will then be some time before the deposites on the surface are all collected; and that it will be not less than twenty years from its first discovery before the whole superficial gold is gathered, and the labor and expense there necessary to penetrate in search of more will require so much labor, enterprise, and capital, that it will be only undertaken by large stock companies; and that the gains will be no more than the ordinary profits on labor and capital, with the chance of some great good fortune and the greater risk of entire loss.

In an estimate I formerly made of the products of the mines for the preceding six months, I may have underrated it, for it is believed that much more gold has been brought to Oregon than I allowed.

I will now estimate the amount to be gathered in future at twelve millions of dollars a year for the remainder of the twenty years; for, although the more easily found deposites will be sooner exhausted, more persons, more machinery, more skill, and more capital will continually be added to meet the difficulty of finding the gold, until the returns cease to compensate the outlay.

This is offered merely as an opinion. The events of the next year may entirely contradict it.

Although the value, extent, and duration of these mines are thus uncertain, the effect their discovery has produced on all public and private interests here is well ascertained and felt in every branch of affairs.

Everywhere else, at least in the civilized world, the cheapest of all those things necessary in the economy of life is mere labor. In most countries there is such a superabundance of it that it scarcely supports its own existence. The smallest talent, acquired skill, or capital, elevates its possessor from this lowest level, increases his gains, and thus enables him to call the labor of others to his aid. The highest rewards of wealth are given to great knowledge, superior abilities, or the most skill and cunning in the accumulation of capital.

Here all is reversed. No employment gives such extravagant gains as gold-finding. And in this the profound scholar, talented professional

man, skilful mechanic, and shrewd trader, are all behind the mere laborer.
The workman who breaks stones on a turnpike, or he who rolls bales of
cotton on the levee, is their superior in all that is requisite to gather gold.
And endurance of hard work, in which he excels, makes him independent of them, while all their superior qualities find no demand for employment, and are of no service to them.

The consequence of this is, that labor here demands the highest rate of
emolument, and cannot be found unless that be paid. No employment
but gold-finding, or those absolutely necessary for its support, is able from
its gains to pay such rates, and is consequently abandoned.

However, there are some who prefer reduced gains with more comforts
and less hardship than are found in digging in the mines. This modifies
the former rate, otherwise the ordinary affairs of society could hardly be
carried on.

As most values are fixed relatively by the rate of wages, it is evident
that here they will be most extravagant. And no deductions can be made
from the uniform and long established prices elsewhere, nor from the prin-
ciples of supply and demand which elsewhere govern.

It is not the plenty of money seeking investment, which, by lowering
its own value in competition, bids up prices to their present high rates;
but the value which is given to labor in one pursuit open to all, abstracts
from every other employment those who are necessary to carry it on, until
they are bribed back by wages approaching their hoped for gains.

These high prices extend into all kinds of trade, but do not indicate
corresponding profits, for these are consumed in the payment of extravagant expenses resulting from high wages. Thus cargoes sold on ship-
board, before they have incurred any expense here, unless they consist of
articles specially in demand in the mines, are generally sold at a loss; but
before they are distributed to the consumer by retail, they are enhanced
ten or twenty fold.

The consequence of this is, that shippers from abroad will generally
lose, and that the final seller of the goods, even at the highest rates, will
retain but a small part of the price. The greater part of this will remain
in the hands of the laborer, or, as but few of this class here are prudent
or economical, in the hands of those who administer to their wants and
their vices.

This state of things cannot be expected to last always; consequently
no important work of improvement will be undertaken by individuals or
associations, since the revenue or profit measured by the future values
can never give even a moderate interest on the sum now to be invested.

Neither can the government in prudence expend money in any work
here not absolutely necessary for the most important ends—such as the
maintenance of its sovereignty, and the integrity of its territories; these
require defences and light-houses to the principal ports, storehouses and
barracks for troops, and a direct and good road from the Atlantic border
to the waters of the Pacific. And these are to be considered not as investments to produce revenue, but rather in the light of a part paid for
salvage to secure the remainder; to be judged of by their necessity, and
not by their cost. I will enter more into detail as to these works of
necessity.

Monterey is an incomplete harbor, being open to the northwest. The
place is a pleasant residence, but has no natural communication with any

great extent of country back of it. Vessels can lie in the roadstead for a long time without much risk from the weather, and troops can be easily landed. A small work commanding the anchorage, and capable of sustaining itself, is all that is necessary now. As such a work will be always commanded by heights in the rear, additional defences will be needed in time.

San Diego is a small but perfectly good harbor, and the entrance is narrow. There is but a small district of country near it, and the ordinary modes of defence for such places would render it too dear a prize for any of its advantages to pay for.

There are small ports in both Territories, opening to limited agricultural districts; and north of San Francisco, to valuable lumber also. But they are not easily accessible, except the most southern ones, on account of the heavy surf which for the most part of the year breaks on the bars at their entrance. Bodega is an exception to this, and is of importance as affording a landing by which the bay of San Francisco may be turned. The mouths of the rivers Umpqua, Rogues, (Requas in some Mexican charts,) Alcea, Yaconna, and Sileetz, in Oregon, south of the Columbia, require to be surveyed, as each furnishes one of the small ports alluded to.

A good harbor is said (but not on certain authority) to exist in Shoal Water bay, about twenty miles north of the mouth of the Columbia river. There is certainly a communication for boats—with a portage of half a mile—from this place to Baker's bay, in the mouth of the river, and this makes the place well worth examination.

The mouth of the Columbia river, though very difficult and dangerous for ordinary navigation, yet, as a natural outlet of a great river, will always be important; and strong works on Cape Disappointment and Point Adams are first necessary. These are seven miles apart, but the intermediate space is filled with shoals. The present channel, after entering equidistant from both, as though direct for Point Adams, turns northward, passes close under the cape, and then curves off eastwardly and south to the point. It is probable that before long a channel will again open from the point out to sea.

Another work on Astoria Point, on the south side of the river, and a similar one on the north side opposite to it, will be hereafter required.

There are probably few harbors anywhere superior to those on the waters of Admiralty inlet, whether for naval or commercial purposes. The entrance to the inlet, and the islands shielding the harbors outside of it, will require strong and defensive fortifications.

The large interests centred in San Francisco bay give it an immediate importance beyond all others, and its defence demands prompt and careful consideration. The straits by which the bay is entered are comparatively narrow, and very bold but short; the tides run with great rapidity, and the winds in the afternoon of the spring and summer, even when it is calm outside, generally blow inward in this gorge with the violence of a gale.

Fortifications on the straits alone will hardly be sufficient to exclude an enemy. An interior line from Sansolita, by Los Angelos and Alcatraces islands, to the second point below the town, will be necessary to make the bay secure. These works would be very extensive, and cost immense sums of money.

The points proper for fortifications are generally bold cliffs; and I would here suggest for consideration, that it will add to the security of the works and to the economy of their construction, that in all the water-batteries the face of the cliff should be scraped to present the desired front and subterranean chambers excavated in the rock behind it for the batteries, and the necessary embrasures opened—the faces and embrasures to be secured by a revetment of hard stone. It must be observed that these cliffs are very high—oftentimes a hundred feet and upwards—and to excavate to the proper level for the work, will cost more than to construct it afterwards.

These cannot be bombarded, and may dispense with protecting fortifications on the heights which command them, except a single work to each system to protect its rear, in which the galleries of communication might centre, and where the troops could be quartered in time of peace.

But the whole system for all points should at once be determined; the works located, and the necessary ground reserved. At present, I would propose the construction of only two small batteries, (to be enlarged hereafter,) one on each side of the Straits of San Francisco, one on Cape Disappointment, and one at the entrance of Admiralty inlet.

But though these defences and other public works are thus limited by the price of labor here, substitutes are within means of the government sufficient for present need.

Though houses cannot be found or built here for the use and comfort of the troops, and for covering the public stores, iron buildings can be constructed in the United States and England, everything prepared and fitted, brought out here and put together by the troops, at a cost less than that required to build frame houses, all the materials being on the ground. But care must be taken that everything necessary to their erection and comfortable use should be sent, for the cost of repairing the omission in California would swallow up all the saving made in other parts.

The houses should be made so as to be subdivided, thus providing for the separation of troops by detachments. Bedsteads, tables, closets and seats could be economically made of the same material, and the whole could be moved without injury.

As a present substitute for those fortifications whose construction is postponed, steamships of war are entirely sufficient. They can be built, fitted out, and manned entirely out of the influence of high prices, and their men can be kept from deserting, while garrisons on the shore have daily opportunities.

Some of the islands north of Admiralty inlet might be made sufficiently secure to establish shops for repairs, from which the men could not get away, or arrangements might be made to fix them at the Sandwich islands.

It will be many years before large armaments from Europe will pass Cape Horn, or that the nations on the Pacific will fit them out to attack us here. In that time circumstances may have permitted us to be perfectly prepared; but for some years a strong fleet of war steamers, with the aid of the few works on shore proposed, will give us the superiority.

If the want of coal be objected, we say that in time of peace it can be procured on Vancouver's island, not far from one of our harbors, the best, nearly, on the Pacific. In time of war, if prepared with such a fleet, we would be more likely to possess the coal than the enemy.

These circumstances receive additional force from the fact that the mail steamers from Panama are not fit for war vessels, and cannot be taken into the service as such, in case of war.

. During the continuance of peace, these war steamers could be permitted to keep up a communication with China, by the Pacific islands, and thus spread and secure our trade and influence in all those seas.

The other work of paramount importance is a secure and rapid communication with the other parts of the Union within our own territory. This, in my opinion, is of more vital consequence than either of the others.

Community of interest, fostered by frequent and easy communication, strong personal connexions and feelings, kept alive by constant intercourse, and the continued interest in public events, when their course can be constantly and accurately followed, beget such an identity of national feeling and patriotism, that under their influence but one heart and one voice would respond throughout the whole land to the call of national honor or duty. But if these people here are to be kept aloof from their Atlantic countrymen, the projects and enterprises of one will be independent and probably adverse to those of the other; the policy that fosters one will injure and offend the other; nations and habits derived from different pursuits and necessities divide, and finally become opposed. If they are not matched in strength, the weaker will cry out, Oppression! and the stronger, Unreasonableness! And finally, when the tie that chafes them becomes too weak to hold them together, they will separate, and from brothers become foes.

The commerce of the Pacific which our territory on its shores is destined to control, is well worth all that can be spent to secure it. And the wealth and power it will bring, may well tempt those who are ambitious of rule to provoke jealousy and indifference to the point of separation, and thus open them a separate field of action.

Nature has placed so much distance and so many obstacles between this and the Atlantic States, that to overcome them is not only beyond the means of ordinary enterprise, but seems too much to be attempted at all. But the very fact of the existence of these difficulties points out the necessity of overcoming them, otherwise they will produce the mournful result I have alluded to. As facility, rapidity, and security of communication are the requisites, a railroad, if practicable, is most appropriate.

The weight of testimony leaves but little doubt, that if the ravines of the Upper Colorado (of the west) can be passed, the road will require less artificial grading, in proportion to its length, than any other over a varied country in the United States.

Whether it be a railroad or a common road, it should cross the Rocky mountains about latitude 38°, reach the valley of Humboldt's river, and follow that direction until it sends off a branch to Oregon by the Willamette valley, and the other by the head of the Sacramento and the valley of that river to San Francisco bay.

It is desirable that the road should cross the mountains as far south as possible to diminish the length of the winter, without going so far as not to reach afterwards the valley of Humboldt's river, by which route one road will serve both for California and Oregon.

If the road were to turn south of the "old Santa Fe trail," where it would avoid the Sierra Nevada, it would approach the Pueblo de los Angelos, and then have to run northward by the valley of San Joaquin to San Francisco bay.

Another road from the United States to Oregon, as well as one from California to the latter, would have to be constructed, each of many hundred miles length. These considerations should decide in favor of the northern route, as common to both Territories, if it be practicable.

None of the important routes in the United States have been constructed at so low an average rate per mile as this will cost; it is the aggregate expense that seems so large.

The most serious objections I see, besides the cost, rise in the uninhabited state of the country through which the road is to pass, and consequent danger from annual burning of the grass; and in the hostility of some of the Indian tribes which may prompt them to steal and destroy.

These objections apply to any road, though not in an equal degree; for all must have many bridges and culverts, the destruction of which would be an entire interruption of travelling, and may in either case be met by establishing colonies or posts, or making small conditional grants for the care of the road running near or through them. It becomes solely a comparison between the value of the road and the cost of making and securing it.

If a railroad be not intended at first, the location and the construction of bridges should admit of its being afterwards graded and laid with rails.

As to that part of the road out of California, the cost will not be enhanced by present circumstances; and for the rest, the importance of the whole must establish the limits of the expense to be incurred in its construction.

While it is important to meet the evils arising from high prices here by such expedients as suggest themselves, it is necessary to consider whether there are no means of removing, or at least modifying, their original cause; for it is not only in public works, but in all public duties, and in all ordinary affairs of life, that the embarrassments alluded to are felt; and as the population of the Territory augments, they must be more serious.

No one can be found to serve in a civil office, the salary of which is totally inadequate to his support. And if the officers, as they are bound to do, confine themselves to the emoluments fixed by law, they will soon abandon their offices, and there can be no administration of the laws of the United States. Such, too, is the temptation to the soldier to desert, either to go to the mines or work for high wages, that no military force can be detached to aid in the execution of the laws with any security that it will return. To counteract in some measure this disposition, I authorized the quartermaster who employed soldiers to work to increase their allowance, so as to make their pay while employed equal to the usual wages of the place. (Division order No. 5.) This was necessary to get the work done, for laborers were not to be hired in sufficient numbers; and it was impossible to conduct work where one was receiving seven dollars a month and the other seven dollars a day. But so soon as the work absolutely pressing is finished, this allowance must cease.

Whatever laws may be passed for this part of the country, requiring the aid of the military in their execution, must be preceded by measures to restrain the desertion of the soldier. Hitherto he has received not only aid and concealment, but encouragement and reward for his crime from the people in the Territory. Laws to punish such conduct *effec-*

ually are necessary: applications to local magistrates or juries of the vicinage would not only be ineffectual, but ridiculed.

While the soldier finds his pay so small in comparison with what others around him are receiving, the situation of the officer is much worse. While his duties and cares are doubled, his few comforts diminished, and his expenses decupled, his emoluments remain the same, and are entirely insufficient. On a journey I made through the mining regions in the mountains, the hired drivers of the pack-mules were receiving more pay than any officer of my staff along; and the drivers of the ox teams we met conveying goods to the mines were receiving more than double all the pay and emoluments of any officer in the division.

Congress are not like to make any increase of pay either of officer or soldier, with the justice of this case. But either the actual cost of those allowances which are now commuted at fixed rates should be given, or, as this provision cannot extend to the soldier, to him and the officers should be given a grant of land, according to rank and time of service here, to be located on any vacant lands in California or Oregon. This would have the advantage of inducing the soldiers to remain by their colors until the end of their term of service, and then to remain and settle in the Territory.

But these temporary alleviations of the evil of certain classes here, are far short of a remedy for the evil itself.

This, as I have before observed, has its origin in the high price of labor, and of other things as caused by that—not, be it remembered, in the general appreciation caused by a superabundance of money. So that, if the price of labor cannot be brought back to its usual standard, all other prices will follow; and then, if the abundance of money has had any effect, it would soon disappear, as through the thousand channels of an active trade it diffused itself through the commercial world.

If honest labor legitimately employed were rewarded in California at ten times its actual rate, it might be proper for private enterprise, or even public policy, to aim at a reduction by introducing or encouraging a fair and open competition, or removing the obstruction to such introduction; but, whatever be the evils, it would never justify any arbitrary restraint on such labor; and, before employing any such restraint, it is as well to inquire into the nature of this labor. Its advantages are from the quantity of gold it collects in the mines. If the mines belong to the government it is not honest to take the gold from them; and digging it on public lands is forbidden by law, and is consequently illegal. And Congress not only have a right, but it is their duty, to restrain this kind of employment as a breach of law and right, as well as a serious evil in all public and private concerns here. These are not Indian lands, as some pretend. Within the limits of the Mexican republic all lands not granted or assigned to its citizens, belonged to the supreme government; and no title was recognised in the Indian tribes that did not recognise in their turn the laws and sovereignty of the republic, in which case lands were assigned for their towns, (pueblos,) and this only among those who were converted. The heathen Indians were considered in the light of enemies, and the lands they occupied granted or otherwise disposed of by Mexican authorities. Those Indians who chose to submit, generally hired and worked on some neighboring farm for their food and clothes; those who did not, retired beyond reach; and throughout the whole of the gold region there are now no Indian claimants, except in the more southern parts, which were too remote hitherto from the settled parts of the country to bring on any collision. Except some that are here in their original state, no Indians are found in the gold region but a few scattered families or small tribes which consider themselves as dependants on the neighboring settlers.

The Indians here are not hunters, and live on roots and berries and the horses and cattle they steal from the whites. Some of the tribes do not pretend to the ownership of anything but the small space occupied by their village, and the grounds around it; but the Mexican government had already got possession of most of the country west of the mountains, and made grants in various parts of it; and, except some such grants, the whole of that district in which gold is found is now public land, not even claimed by any Indian tribe.

Even for that portion still at the south occupied by Indians, it will be difficult to apply our laws in their favor. They are, like the North American Indians, treacherous, cruel, and dishonest; but far inferior in every intellectual and moral quality to any we have been accustomed to deal with.

These mines are, then, on the public lands, and under the protection of the laws forbidding trespasses on public lands, and reserving mines for the public use.

If there were the necessary judicial authority here, and the force adequate to execute its decrees, a stop could be put at once to the digging of gold, and to all the evils which flow from it. But it is not probably the policy of the government nor the wish of the people of the United States to buy the partial advantage here by the loss of those resulting from the activity given to general trade by the diffusion of so much of the precious metal.

I do not conceive that it would be desirable to have the mines worked for the benefit of the public treasury. To do that would require an army of officers and inferior agents, all with high salaries, and with opportunities and temptations for corruption too strong for ordinary human nature. The whole population would be put in opposition to the government array, and violent collisions would lead even to bloodshed.

If the government shall desire revenue enough to pay the expense of executing the laws passed on the subject, it is the most that should be proposed. The advantage which the whole country will derive directly from the opening of the mines, and the indirect advantage to the treasury from augmented commerce, will, in my opinion, more than compensate for any outlay it has made or may make.

I would propose, then, to open the mines to the public, but under such conditions and restraints as will keep in the hands of Congress the right to modify and improve its plans, and as will tend to remove or diminish the evils now experienced here. To survey and sell the lands will be too tedious. In the mean time, all the difficulties now experienced would continue; and, after the lands are for sale, contests for individual rights would keep the whole country in litigation for years to come. Any delay in the remedy will render it almost useless. In lieu of a sale or a lease, which would also require time and a survey, I would propose the following plan:

Divide the gold regions into four districts. Grant licenses to dig or
Ex.—7

seek gold in a particular district, anywhere, not within a certain distance of any other licensed person, for a certain time, on the payment of a fixed sum. No person to be licensed but a citizen of the United States. This would leave branches of labor and industry, with all their advantages, still open to foreigners. Every person employing or harboring a deserter, to forfeit his license and gold, and be subject to a fine, and not allowed to procure another license. Soldiers, at the end of an honorable term of service in California or Oregon, to be entitled to a year's license gratis.

A "tribunal of mines," consisting of a president and four associates, to issue the licenses, and try all disputes about mining, first, before one of the associates in the district, with an appeal to the president and two others, at regular terms.

The "tribunal," having authority alone, or in connexion with a jury selected or drawn by lot, to make ordinances and regulations for the government of the mines—an authority similar to that exercised in relation to roads and prisons in many of the States by county courts.

The laws of Spain relative to mines in her colonies contain many excellent details of a system such as this.

The mines and the lands they are on being the property of the United States, they have a right to exclude all from them, and, consequently, the right to admit them, on such conditions as they may choose to prescribe.

Where, as in the United States generally, the right of mining belongs to the owner of the soil, it requires no additional regulation, as it is governed by the same laws as other property; and if these lands were sold, there would soon be no difficulty, for the owners would each regulate his own property; and any attempts to enforce restraints on the use of the public lands by the ordinary tribunals of the State or Territory, where the whole population is engaged in the offence, is vain: while, by this plan, the general right of seeking for gold is limited, and produces a revenue, and the interest of those licensed is engaged to prevent infractions of the law, and to support the authorities appointed to maintain it.

If the laws could be enforced by local tribunals, it would be unjust to tax the people for the extraordinary expense arising solely from this branch of business. While this system is in progress, the lands can be surveyed and examined. Those that are destitute of gold first sold; then those that become exhausted; and, finally, the whole, as the surveys and examinations are completed, according to the demand at the time, when all these special laws may be repealed.

The crime of desertion among the troops is not only an evil at all times to be guarded against, but circumstances here render it peculiarly desirable to find a remedy.

Thousands of people from foreign countries have flocked here to gather gold and carry it away. They have no interest in the future welfare of the country, and, if they thought themselves strong enough to succeed, would join together to oppose the execution of any laws restraining their illegal labor. And it thus becomes very necessary to keep the military corps stationed here to their fullest complement.

It is very much to be regretted that the punishment of death was taken from the law against desertion. Not that it is desirable that it should be inflicted often, or even in but few cases. And I believe, while it remained in the law, there was no instance of the infliction of that punishment in time of peace. But it should have been left to the discretion of the court, tempered by the clemency of the Executive, to measure out the punishment according to the aggravation and exigencies of the case. The offence here is a breach of contract and of oath; but the penalty is not so much to punish that as a crime, but to maintain at all risks the military force entire, so many lives and so many interests being often depending on that. Still the greater evil arising from the change is the idea which it has given rise to the soldier—not that modern humanity has moderated the severity of the punishment, but that desertion has been discovered to be less of a crime than it was formerly; that the obligations of honor, duty, and oath are diminishing, and to desert now is much less disgraceful, as it is made less dangerous. But it must be acknowledged that in the present state of things in California, severe enactments, though they might strengthen the hold of the service on the soldier, could but seldom be put in execution, such is the difficulty of detecting, arresting, and detaining deserters from the aid afforded them by citizens.

The additional or severer penalties should in a measure be self-operative, so as not to be escaped by absence or concealment.

To be *reported* as a deserter, should, *ipso facto*, cause the forfeiture of all the delinquent's property, and he should be incapable of acquiring, possessing, or transferring anything, or of making suit in court, so that he could have no hope of acquiring permanently anything by mining; and those who dealt with him should gain nothing, and lose whatever they bought from him, or the price of what they had sold him; imposing on them, in all cases, the burden of showing that the person, if so reported, was not a deserter. Reserving to any one who thought himself unjustly reported, the right of demanding a trial on the charge; and, as many have already deserted, continuing absent beyond a fixed day after the passage of the law to be created an offence, and punishable as desertion under the new enactments.

As the military force is to be called on in certain cases to prevent trespasses on public lands; and as some of these lands are to be reserved for military uses; and as population is flowing into these Territories desirous of acquiring lands, it is highly important that provision should be made at once for ascertaining what is private property, or belonging to corporations, and for securing that which belongs to the government.

Both California and Oregon, though differently, are peculiarly situated in regard to this subject. Under the Spanish and Mexican laws no general survey of the public domain was made. Application for land, describing its situation, was made to the governor. The local authorities were on this ordered to examine whether it was unoccupied, or necessary for the public use, and the character of the applicant; on their report and certificate further steps were taken, until the grant was complete by a survey and the confirmation of the departmental legislature and the supreme government. Sometimes the grant came direct from the latter as a special favor, or in payment of claims or debts.

There was but little land cultivated in the Territory except at the missions, and the only commerce was in hides and tallow; to maintain this only large herds were necessary, and these required extensive pasture, being never housed, and finding their food at one season in one place, and in another at another. The inhabitants were few and widely scattered. Their mode of life and employment is the origin of the large grants of land in this Territory, varying from one to eleven square leagues, according to the number of cattle to be pastured. A Mexican league is about two and five-eighths of our mile, (5,000 varas, each of 3 burgos feet,) and a square league contains a little over 4,000 acres.

So few indeed were the inhabitants in proportion to the extent of lands, and so little danger of collision was there, that no enclosures marked the boundaries of their property; and while there was so much vacant land unclaimed, no one coveted his neighbor's. From this arose great negligence, both in ascertaining their exact bounds and in pursuing all the forms pointed out by law for the completion of their titles.

Many on their first petition and order finding no alcalde at hand to make the requisite examination and certificate, took possession; and never having it disputed, have remained to this day without taking any farther steps. Some have gone further, and some have completed all the forms. But the different stages of title were all on a level in public estimation until our acquisition of the country enhanced the value of the lands and induced an examination of the records.

This looseness in regard to titles has induced the purchase by speculators of anything approaching to any of these forms, and much of the public domain will be claimed on the most insufficient grounds; while on the other hand many honest proprietors, misled by their own ignorance and the negligence of their own authorities, would, under a strict application of law, be deprived of much of the property they have possessed and lived on for years. Great frauds were perpetrated under the latter day of Mexican rule, and many after it ended, though their apparent date is before.

Towns (pueblos) were created by law, or by virtue of its provisions. A large tract was set aside, in the centre of which lots for dwellings were laid off. These were granted by the ayuntamiento or town council to persons becoming residents, on condition of their building on them. Their grant gave them a right of common on the exterior grounds; but any one having this right might demand the exclusive use of a share of it, which he might enclose and cultivate; but it remained a part of the common property of the town, and reverted to it when he ceased to use it.

This species of property has been disposed of without proper authority, and not in the way ordained by law; and it is the interest of the government to learn what title or right remains in the lands thus disposed of.

The first settlement of the country was made by "missionaries." They were authorized to gather the Indians around them, to convert, instruct, and civilize them, and to cultivate the land around them for their support. At one period, large tracts were reclaimed for this use, and much revenue produced from the labor of the Indians. No property belonged to the mission, except the church, parsonage, and adjacent buildings, and a small piece of ground around them; and the title to this was really in the religious order or society in Mexico to which the missionary belonged—was "ecclesiastical property," and, as such, could only be alienated for certain reasons, and under certain forms. The "mission" was not a corporation, but merely the agent of an ecclesiastical one in Mexico. The Mexican government has resumed all the lands, and taken the missions as national property—an act, the legality of which is disputed by the missionaries. But certain portions of the land are assigned to the "mission Indians" for their support, under the administration of the priests—the former not being endowed with full civil rights. Some of this property is claimed by subsequent grants from the government, some by purchase from the Indians, and the priests have sold some—though neither of the two last can have authority to sell, under the Mexican laws.

All these irregularities, which increase every day, bring great confusion; and finally, when ignorant but honest buyers are involved, strict right cannot be done without great loss to those who least deserve to suffer it.

In the mean time, many industrious emigrants, desirous of buying homes, are flocking into the country, and find nothing for sale the title to which is sure.

It cannot be expected that legislation will at once attack and cure these irregularities. A careful research into the laws and records, both here and in Mexico, is first necessary. But there needs here at once a board of commissioners, or some similar authority, for examining all the titles held against the United States. Those grants, no matter how large, which were good under the Mexican laws, are good now; but those which were not complete should be arranged in three classes: 1st, those merely wanting the confirmation of the supreme government; 2d, those consisting of the first petition and order; and 3d, mere possession and use—confirming to each on the principle of the confirmation of grants in West Florida.

There is no injustice in thus diminishing the grant, for no one has a title in law until it is all completed; and the lesser quantity confirmed under our government is actually worth more than the larger under the Mexican was: so that we increase the value, though we diminish the quantity. But the most important end is to give security to title as soon as possible; and to do this, a loss of land would be preferable to a postponement of the matter.

Oregon is in a situation entirely different, but still embarrassing. There are no legal titles in the country, except "the possessory rights of the Hudson's Bay Company, and of all British subjects who may be already (June 15, 1846) in the occupation of land or other property lawfully acquired within the said Territory," and "the farms, lands, and other property of every description belonging to the Puget's Sound Agricultural Company, on the north side of the Columbia river," (treaty of Washington,) which are recognised and confirmed by the treaty of Washington.

Congress was prevented from surveying, selling, or granting the lands in Oregon by the policy which had left undetermined so long the respective rights of Great Britain and the United States therein. And the surest way to have the question determined in favor of the United States was to have the country settled by emigrants from herself, who would not only secure the possession of the country to us, but make it little desirable to others.

The people who accomplished this underwent incredible hardships, and have suffered losses from petty Indian depredations that far outvalue the price of the land. They left nothing undone on their part to get the opportunity of acquiring legally a homestead here; but matters in which they had no concern postponed the usual legislation on this point, and the acts of their own informal legislatures on this point were left unnoticed by Congress and unrepealed. Under these circumstances, the settlers "claim" the section of land which they have occupied.

I am certain that the expense and loss of their journey out here, and the difficulties they have met since their arrival, would be poorly paid by the price of the land they claim. But they have a merit beyond this, in the service they have rendered to the country by establishing its title here, and in the fruitless efforts they have made to draw the attention of Congress to this subject long since.

But the merit of all is not equal: 1st, some have *bona fide* cleared, settled, and improved their farms, on which they now live, supporting their families; 2d, others have just brought themselves within their own laws respecting "claims," solely for the acquisition of the land; and 3d, others have seized on places they suppose desirable to the government, and bring themselves just within the pre-emption laws for other districts, merely to acquire the property, and sell it afterwards to government; or those who intend to speculate again on it.

The first is surely entitled to the land which has already cost him so much; but the others should only get what a strict construction of the laws shall give them.

I would especially urge that it be recommended to provide:

First. That no incomplete title shall cover a tract that government may want for public use.

Second. That, before lands are offered for sale, the War, Treasury, and Navy Departments be called upon to know what is necessary for military and naval purposes, custom-houses, and light-houses.

Third. That if any one, by incomplete title in California, or claim in Oregon, or pre-emption right in either, acquire any right to a tract of land necessary for the public, an equal quantity shall be given elsewhere; and improvements made before the passage of the law shall in certain cases be appraised and paid for.

Claims exist or are prepared for many points in both Territories that can be of no use to the owners but to sell afterwards to government.

It seems absurd that the United States, gratuitously, or for a nominal sum, should to-day divest itself of property, only to buy it back to-morrow at an enormous price.

By at once marking off all that could be possibly required in future, and allowing no private title to cover it that was not perfect of itself, contesting claimants would be excluded, and selections could afterwards be made by the proper agents of that absolutely wanted, and the rest restored or put up for sale.

In the Atlantic States this difficulty never has arisen, for the private ownership was generally older than the government; and, except in Louisiana and Florida, there were no public lands. But here is a coast over a thousand miles long, with many important harbors, and the best lands lying near the coast; and already many of the positions necessary for defence are subjected to private claims, solely with a view of selling them afterwards to government. It would be well worth while to learn whether by the laws of Mexico now, as I know it was before her independence, concessions from the government were not liable to be entered on and reoccupied for necessary purposes of public use, reverting to the original grantee when the government abandoned them. And if such be the case now, let the owner of even a complete title be compensated by an equal quantity elsewhere, if he do not choose to await the abandonment of the property by the government.

I hope, for the security of troops and stores on this coast, where they must generally be transported by sea, and for the interest of commerce generally, that appropriations may be made this year for the construction and erection of light-houses for the harbors most likely to be used. So much time is lost in communicating between this coast and the seat of government, that some years may elapse if they wait for detailed reports on the subject. All this coast has bold and prominent points, that at once indicate themselves as the proper positions for light-houses.

There should be one on Protection island, near the entrance of Admiralty inlet.

One on Cape Flattery, at the entrance of the Straits of Fuca.

One on Cape Disappointment, at the mouth of the Columbia.

One on Punta de los Reyes, a promontory north of San Francisco bay, which vessels coming from the northward and westward generally make, being to windward of the port.

One, the most important of all, on the Farallones Rocks, directly outside of the entrance to San Francisco bay. These are rock islets, with plenty of room for the buildings, and a small cove for landing in boats.

One in the entrance of the straits on the right-hand point going into San Francisco.

One on Pine Point, (Punta Pinos,) near Monterey.

One on each of the islands of Santa Cruz and Santa Catalina, lying south of the "Channel" of Santa Barbara.

And one at San Diego.

With the exception of San Diego and the two on the Straits of Fuca, which I expect to visit in a few days, I have seen all these positions; and as in all, except San Diego, the foundations will be rock, the light-house frame and the necessary buildings may all be made of iron, in Europe or the United States, and be put up here at once, everything being provided. They should all have bells—some two, for distinction—to sound in the fog.

The length of voyages would be much shortened if vessels could approach the coast in a fog with the certainty of meeting with some indications of their exact position.

A subject of great importance to the public interests here, and particularly to the military branch of them, is the acquisition of the rights and property reserved and guarantied, by the treaty of 1846, to the "Hudson's Bay Company," and the "Puget's Sound Agricultural Company." The "Hudson's Bay Company" is a corporation chartered, for the purpose of trading, by Charles the Second, on the 2d of May, 1670. The lands on all the waters of Hudson's bay, called "Rupert's land," were granted by the same royal charter which incorporated the company. In 1787, the Northwest Company, in Canada, established a fur trade in the British northwest territory, which, by surrounding the outer posts of the Hudson's Bay Company, intercepted the trade of many of the Indians who frequented them.

After such rivalry, contest, and finally bloodshed, tending to the gradual ruin of the Indians and their fur hunting, the two companies combined; and, under the provisions of an act of Parliament of the first years of "George the Fourth's" reign, on the 6th of December, 1821, the association was licensed for the exclusive trade of all the British possessions in North America, not organized in any provinces or included in "Rupert's land." Saving the rights of American citizens on this coast under the convention for joint trade, this license was for 21 years.

Under this license, posts were established in Oregon; and Fort Vancouver, the principal one, was begun in 1824. In 1838 this association was dissolved, by the Hudson's Bay Company buying out the interest of the others, and admitting some of them as members of the company; the license was consequently surrendered, and, on the 30th of May, 1838, a new license for 21 years was received by the Hudson's Bay Company for the exclusive trade of the same region, reserving always the rights of American citizens, under the convention before mentioned. This license was unexpired when the treaty of Washington was concluded, and it found this territory open to the trade of the British with the Indians by our consent; and this trade, by the special grant of its own government, enjoyed and exercised by the Hudson's Bay Company a perpetual corporation, capable of holding real estate; which had, in the exercise of the right recognised by both countries, built houses for conducting their trade and sheltering their goods and servants—defended them by stockades—cultivated farms and occupied pastures for providing food not only for these posts, but for the distant ones in a climate and soil incapable of yielding it.

The treaty recognised the right of the company to be here; because it secures not only to it, but to all British subjects trading within it, the free and open navigation of the Columbia river from our mutual boundary to its mouth, and it recognised its right to have the houses, lands, and other things, by guarantying that its "possessory rights" shall be respected in the future appropriation of the territory. In other words, that they shall remain undisturbed, either by public or private encroachment, in all the property or rights they possessed at the date of the treaty, whether houses, lands, ways, pastures, cutting wood for mill or fuel, or connected with either.

Their rights under their license were of trading exclusively with the Indians, as to British subjects. These expire with their license, in 1859; but the treaty makes no reference to their license, and grants not only their possessory rights, but the free and open navigation of the river to the company, without other limitation than exists in their capacity to hold, and their charter makes them a corporation "forever."

Their license to trade in Oregon cannot be renewed, of course, but the right of free navigation may be more important to them for the trade north of our limits than within them.

The farms, lands, and other property of every description, belonging to the Puget's Sound Agricultural Company on the north side of the Columbia river, are specially confirmed to it by an article of the treaty.

The two companies, although existing separately, are, I believe, connected by some interests, and the property of both is for sale in one transaction.

It is undoubtedly the interest of the government to own nearly every portion of the real property offered. Some parts of it are indispensable. But the extraordinary privileges given are of far more importance than the value of ten times the property would be.

If a grant of the "free and open navigation" of the river be made a special article of treaty, it certainly must mean to convey something beyond what would have accrued to the parties by the mere operation of law; it must be something more than, and beyond, any right which they would have had without the treaty; and if it be so, the actual loss of revenue alone, and the expense of preventing smuggling when the country is thickly settled, independent of the jealousies, collisions, and claims for indemnity which will arise, will more than overbalance any price which may be ásked. But political reasons of expediency, connected with the undivided control of commerce within our own waters, are alone sufficient to bring this subject quickly to a conclusion.

As to the property desirable as a purchase at Coleville, Fort Vancouver, Puget's sound, and Cowlitz river, are farms in cultivation, furnishing provisions not only for the company's posts, but from Vancouver and Cowlitz are supplies shipped to the Russian company's post on the northern coast.

In all these establishments, besides the farms enclosed and cultivated, there are large ranges for cattle; and if the extent covered in this way be the rule that would obtain in Texas, New Mexico, California, here or any other pastoral country, the possessory rights of these companies are very large.

But if some of their rights are not well defined, and the extent of their claims may be disputed, it is rather an additional evil, only to be avoided by a purchase; for the government prefers to the acquisition of acres, putting an end to uncertainty of title, avoiding litigation, and cultivating peace and good will within its borders, so that the resources of its newly-established territories may be soonest developed.

The positions selected by the company in what was at first a hostile country, in their relations to the Indian tribes, they meant to control; their topographical features, means of commmunication by land and water, facilities for subsistence and connexion with each other, are precisely those desirable for our military posts and depots.

Fort Vancouver has every requisite for the principal garrison and depot, and centre of all military concerns of the department for a long time. There are buildings sufficient for all the stores of the quartermaster's, commissary's, and ordnance departments, for barracks for the men, for hospitals, and, with some additions, for officers' quarters and stables.

The farms are in good cultivation. There are a saw-mill and grist-mill on the ground, which would be now of great use. Such is the demand for lumber that it is now hard to be procured; and the flour which might be ground here would be a great aid to the commissary's supplies, for the flour shipped around the tropics, twice passed the tropics, is much of it sour.

The machinery for a saw-mill, to cut very large logs, with a planing, tonguing and grooving, and shingle-cutting machine, and other adjuncts now common in the United States, should be sent out, even if this mill be purchased.

The cost of all the buildings about this establishment is estimated by the quartermaster to have been beyond seventy thousand dollars when they were built: now, they would cost from five to eight times as much, and, from want of labor, could not be finished under two years. To build new establishments of equal extent, in the style in which government works are to be constructed, would cost from five to eight hundred thousand dollars, and might in some years be found out of place.

These will last, with care, probably from ten to fifteen years; and then, with extensive repairs in the less permanent parts, some time longer. At that time it will be advisable to move the establishments.

The value of the buildings now estimated by former prices, and considered as already old, is not one-tenth of what it would cost to build the same anew.

I feel anxious to learn that this matter has been arranged one way or other, because, in May, many storehouses and other buildings are to be provided. If these are bought, they will nearly suffice: if they are not, others 'are to be built. Should a purchase be made, it is important to get possession of the saw-mill immediately.

I however put no advantages of conveniency or economy in comparison with the necessity of resuming as soon as possible the privileges and concessions made to the company.

In another view of the case, the people of the United States owe the company a large debt of gratitude for their liberal and unvarying kindness to the citizens who have emigrated there. Many of the first comers owe their lives to the generosity of the company, who opened its stores to the wants of a destitute population. And it is now entitled to more security from annoyances, intrusion, and tresspass, than the administration of the laws in a new territory is able to afford it.

The interest of both parties points out the propriety of getting at once the unlimited control of the whole country.

If the purchase be made, I would respectfully advise that it should particularly describe the lands, and that these be reserved from sale, claim, or pre-emption, and should be in time sold as may be found convenient; otherwise the whole, houses, mills, and all, will be snatched up by squatters, who are already intruding everywhere, nothwithstanding all the efforts of the company's officers to maintain their property by legal means.

If the arrangements cannot be made this winter, appropriations will be necessary for the necessary buildings elsewhere, for there are no others to be had in the Territory.

California formerly, and Oregon until this year, produced an abundance of wheat, not only for consumption but for shipment, and there are in both Territories now abundant supplies of fresh meat. But now the cultivation of the earth is nearly abandoned, and the herds, particularly in California, are neglected. Men cannot be procured to drive them to market, or to slaughter and prepare the cattle.

California produced barley, and Oregon oats, but these grains are now and will hereafter be scarce, from the reason mentioned.

Fresh meat can, however, yet be procured but at increasing prices, in both departments.

Flour and barley can be got from Chili; coffee and sugar from Mexico, Central America, and the Sandwich Islands; salt meats and the "small rations" must still come from the United States.

Major D. H. Vinton, assistant quartermaster, has accompanied me over the greater part of both Territories that I have visited, and is provided with minute memoranda, not only of the nature of the country, but of the necessities of the service and the convenience of supplies in his department.

Major R. B. Lee, commissary of subsistence, has also carefully examined into the resources of both departments for the supply of his branch of the service. Both of these officers are to return immediately to the United States, under the orders which brought them here. Their reports to the chiefs of their respective departments will contain a great many details on the service in each, to which reference can be made.

Since I began this report, I have received information of the arrival of

Lieutenant Hawkins at Fort Hall, and that the advance of the train with a year's provisions from the United States was in; so that Lieutenant Colonel Porter will remain at that post, and in the spring two companies of mounted riflemen will occupy the Dalles, and will have to construct some additional houses there.

I must advise the general commanding in chief that there is no regular mail to this department, (Oregon.) The company which contracted to bring it from Panama have never sent a boat beyond San Francisco, and it is very seldom a mail is sent from the latter place here. Contracts to carry it in the interior cannot be made, for they are to be paid out of the "proceeds of the post office in Oregon and California;" and as no one knows what that is, no one will undertake it. It is also hard to find any one to serve as postmaster for the insufficient compensation. Occasionally mails are sent by transient vessels from San Francisco, but in the winter they will be long delayed getting in at the bar, and those that are sent getting out. In California the mail comes with some regularity to San Francisco from the United States, but the boats do not stop always at Monterey and San Diego, and it is hard to maintain a communication with the latter place by land. From the difficulty of communicating in the interior, the departmental return of California had only been made to May, and that was received by me in August: from an error in its calculation it had to be sent back.

The transmission of official communications between this and Washington will of course be very irregular. The last that I have from the headquarters of the army are of the date of June.

Since I left San Francisco, on the 1st of September, I have received no intelligence from California. In a newspaper in the possession of an officer of the Hudson's Bay Company, I perceive that the convention had met and organized. This was the 3d or 5th of September.

This report was intended to be sent by the steamer that was to have left San Francisco the 1st of October, and would then have arrived in New York about the 10th of November; but being obliged to come to Oregon in a sailing-vessel, I lost the whole month of September, and no vessel with a mail has since got out of the river. This will go by a transport taking down lumber, which is now (November 7) ready to go. She may be detained an indefinite length of time at the bar, so that much later intelligence will be received at the headquarters in New York from California than I can send, or, indeed, than any I have here.

November 12.—Paymaster Reynolds arrived here to-day from Fort Hall. He reports that some scurvy had appeared among the men there, who had no vegetables. They had sent to the Salt Lake for some, but were not certain they could be procured. Major Reynolds was forty-four days from that post here, and had snow in the upper country; so all communication is stopped until spring.

I have neglected to advert to one point, which I will now touch upon. The horses in California and Oregon are none of them fit for dragoons; they are too small to carry a man with his equipments. They are, indeed, active under a light weight, but not more than our blood-horse, properly trained, is under a heavier one, and have not his endurance. Their only advantage is their habit of feeding, which requires no forage to be carried for them when there is pasture; but our horses acquire this habit.

The rifles lost horses on their route; many were stolen by teamsters and soldiers deserting, who of course took the best. But many were broken down from being put at once on a march before either horse or man could be trained.

Every year a certain number should be bought, and sent early in the summer to the most westerly cavalry post, there exercised and trained during the fall and winter: one hundred sent every spring, halted in the fall, and remain the winter at the post this side of the Rocky mountains, and sent in here to the troops as soon as the grass serves in the spring. Now, after a long journey they arrive on the Columbia when the pasturing and water, except in the river, are dried up, and the whole country burnt off. Detachments of troops or recruits could bring them out.

For merely moving the men through the country, the horses of the Cayuse Indians will do. To buy them, articles of Indian trade, such as blankets, beads, hatchets, &c., should be sent to the quartermaster. Then the detachment sent to serve in their country would go on foot, buy horses, and mount themselves, and in turn be relieved by another dismounted detachment.

I neglected, too, to mention that I had seen no granite in California or here. There is a very hard sienite on the south fork of the American river, near Mormon island; it is a very hard stone. For revetment for the fortifications on the coast it would cost too much. Granite, quarried and cut to any face or shape, could be procured cheaper from China.

Your obedient servant,

PERSIFOR F. SMITH,
Brevet Major General United States Army.

Brevet Lieut. Col. W. G. Freeman,
Assistant Adjutant General.

Lieutenant Theodore Talbot's report to General Smith.

Fort Vancouver, Oregon,
October 5, 1849.

Sir: In pursuance of your orders enclosing instructions from Division Headquarters, of June 24, 1849, and directing me to carry out that portion of them relative to an examination of the Alcea river and the country adjacent, I proceeded from Fort Vancouver to Oregon City by water, on the 14th of August, with a detachment, consisting of a sergeant and nine men. I was delayed here some days, in consequence of the difficulty in procuring saddles, bridles, pack-saddles, and other requisites for the expedition—the great number of parties constantly leaving for California having completely drained this place of those equipments. I engaged here Joaquin Umphraville, an expert French voyageur, as my interpreter, and to take especial care of the pack-horses. The sergeant of my detachment, being taken seriously sick, was left, with orders to return to Vancouver.

Having completed our preparations, we started from Oregon City on the 20th of August, travelling eighteen miles up the eastern side of the Willamette to Champoeg, an old French settlement on the banks of that river. The next day we crossed the river at a ferry three miles above

Champoeg, and proceeded by easy marches up the valley of the Willamette, crossing the "Yam Hill," the "Richreol," (a corruption of the old French name "La Creole,") and the "Little Luckiamute"—streams all tributary to the Willamette, and taking their rise in the Coast range of mountains.

The country through which we passed was moderately rolling, about one-third being covered with timber, the rest prairie or open land. The forest consists principally of white and live oak, and different species of cedar, pine, and fir. The soil of the bottom lands is a brownish loam mixed with blue clay; that of the uplands is loose and gravelly. Claims are located and more or less improved on nearly all the advantageous sites for cultivation; but at present evince general neglect, many of the farms having been altogether abandoned by their owners for the more rapid acquisition of wealth in the mines of California.

On the 24th we reached "King's Valley," a pretty plain, some six miles in length, and from one to two miles in width, lying immediately at the foot of the coast range, and separated to the eastward from the main valley of the Willamette by a line of steep hills. It is watered by a stream called the "Big Luckiamute." Four families are settled here, and have well-improved farms. The distance from Oregon City is estimated at sixty-five miles. From the best information which I could obtain, I selected this as the most favorable point at which to pass the Coast mountain.

August 25.—Crossing the Luckiamute, which takes its rise further north, we took a nearly west course, following a small Indian trail, which led us over a succession of high, steep ridges, running nearly at right angles to our course, and covered with forests of pine and fir, and a dense undergrowth of brushwood and fern. We crossed several small streams, the headwaters of Mary's river, a tributary to the Willamette. The mountains were enveloped with such a dense mass of smoke, occasioned by some large fires to the south of us, that we could see but little of the surrounding country. These fires are of frequent occurrence in the forests of Oregon, raging with violence for months, until quelled by the continued rains of the winter season. We met on the road a small party of Cliketat Indians returning to the Willamette from a hunting expedition. The proper range of these Indians is on the east side of the Cascade mountains; but they have gradually encroached upon the hunting-grounds of the tribes to the west of them, until they have reached the very ocean itself. Within a few years past they have cut the only two trails (one of which we are now travelling) that cross the mountains between the Willamette valley and the coast. I obtained from them a good deal of information with regard to the part of the country over which I wished to travel. We made our camp on a small stream, walled in on either side by steep mountain ridges. The horses that I had been furnished with were nearly all in very poor condition, and entirely unfit for any rough service, as to-day's travel proved. Two of them gave out completely, and were left behind; and four others were with difficulty brought up to camp, although we had only come fourteen miles.

August 26.—Our road to-day, like that of yesterday, was full of steep ascents and yet more precipitous declivities, and much obstructed by fallen trees and thick brush. We passed through one tract of burnt forest several miles in extent, where the little trail which we followed, indifferent at best, was often completely broken up, and we were compelled to have recourse to our axes to make a way through the heaps of charred logs. We descended, after a toilsome day's journey, into a grassy valley, about half a mile in length, watered by a fork of the Celeetz river, in which we encamped, having made nine miles.

August 27.—We travelled down this stream, struggling through the dense willow and cherry thickets which line its banks. Two miles below our camp of last night we struck the main fork of the Celeetz river, flowing from the NE. It is about forty feet wide, with an average depth of three feet; the bottom rocky—large boulder stones in many places breaking the rapid current. Crossing it, we ascended the bank into a handsome prairie, extending several miles along the north side of the river, which from the junction of its forks takes a nearly west course. The soil of the river bottom is very rich; grass growing most luxuriantly where not completely choked up by the fern—this plant usurping possession of nearly every open spot of ground. It grows here from eight to ten feet in height, and is quite a serious impediment to travel. We encamped in an open prairie bottom about a mile long and half a mile in width, just where the river, changing its course, makes an abrupt bend to the north. We were surrounded on all sides by tall forests of pine, fir, spruce, hemlock, &c., which gave quite a sombre appearance to this sequestered valley. I had the pleasure of meeting here his Excellency Governor Lane, and two other gentlemen who had designed accompanying my party; but we had missed each other in the Willamette valley, and obtaining a Cliketat Indian as a guide, they had come on in advance, and were now returning.

Having been informed that coal had been found near here by a party of whites who had visited the Celeetz about a year since, I devoted a day to an examination of this locality. We saw indications of coal at several places in the north bank of the river, and at length, after considerable search, found a seam four inches in thickness just below the surface of the water. It had a dip of forty degrees to the north, and was thirty feet below the top of the bank, lying under a bed of shale or slaty clay, sixteen feet thick, and fourteen feet of loose gravel and surface soil. In the superlying shale were many discontinuous seams or streaks of coal from one-fourth to half an inch in width. Specimens of this coal have been submitted to the inspection of practical miners and others, who pronounce it to be anthracite, and of good quality. The lower strata along the river banks are generally concealed from view by the masses of rubbish which have fallen from above, and by a tangled growth of briars and thick brush, which it would require much time and labor to remove. There is but little doubt, however, that larger seams of coal than the one found by us must exist in this vicinity, probably near the same depth below the surface.

The distance from the bend of the Celeetz to King's valley is thirty-four miles. The Indians say that a canoe can descend from here to the ocean in two days, but that the river is full of rocks and rapids and the navigation dangerous. There are no Indians residing permanently on this river, and no trails going further down; the one which we have followed thus far crossing the river here and striking south.

August 29.—Parting company with Governor Lane, who returned to the Willamette, we forded the Celeetz at some rapids, and, travelling four miles through rolling hills, ascended a steep and heavily timbered mountain. I saw here pines eight and ten feet in diameter; the alder also grows to considerable size, many trees being eighteen inches in diameter. The trail often wound along the edge of lofty precipices, where one false step would have plunged us down hundreds of feet into the rocky ravines below. The dense fog, however, concealed from us the full extent of the danger. Descending the mountain, we found ourselves on the shores of the Yacona bay, where we encamped, having made twelve miles from the Celeetz.

Riding a mile down the shore of the bay with my interpreter, we came to a small Indian village, whose occupants received us very kindly. They call themselves Yaconas. The Indians residing on the Celeetz, Yacona, and Alcea bays, all speak the same language and belong to the same nation; but each bay has its respective chief. There are about 80 of them, all told, living on this bay. They are generally well formed, intelligent, and of healthy appearance, apparently not being subject to those eruptive diseases of the skin which prevail so extensively among some of the tribes on the Columbia. Most of them talk the Chenook jargon, a singular medley of corrupted English, French, and Chenook words, spoken by the different Indian tribes of this coast in their intercourse with each other and with the whites, somewhat as the French language is used among the polished nations of Europe. The Yaconas subsist principally on fish, crabs, clams, and roots, occasionally hunting the elk in the neighboring mountains. They do not possess any horses, and have had but little intercourse with the whites; neither the chief nor any of his people had ever visited the Willamette valley. Having given them some presents, I explained to them the desire of the Great American Chief to establish and preserve friendly relations with all the Indian tribes.

August 30.—The grass being very scant on the border of the bay, I sent the horses back two miles to a little grassy valley in the mountains, which we had passed yesterday. Hiring a canoe, and five Indians to manage it, I went down the bay to its outlet, (into the ocean,) which is about three-fourths of a mile wide. On the north side of the entrance are high yellow sandstone bluffs covered with fir trees; on the south side a cape of low sandy hills, with clumps of dwarf pines. I sounded the channel, with which the Indians are perfectly acquainted, from the entrance to the head of the bay, a distance of about four miles. The depth of water ranged from four to seven fathoms; general width of the channel forty to fifty yards. For a mile and a half from the entrance, the channel keeps near the north shore of the bay. There are two sand-bars about half a mile from the entrance, but they do not interfere with the channel. The land on the north of the bay is all high; on the south it is much lower, both sides being covered with forests of fir, spruce, hemlock, cedar, &c. The bay varies from one to two miles in width; a large portion of the upper part is very shallow, being left nearly dry by the receding tide. I ascended the Yacona river five miles. The average depth of water in its channel was twenty-four feet. The river is bordered by very steep hills covered with a forest of evergreens. The Indians say there are no trails leading up this river, that the country is very broken, and the forest impenetrable. We returned to camp in the evening, half benumbed with cold, the day having been most unseasonably chilly and very misty, much more resembling midwinter than the height of summer.

August 31.—Crossing the Yacona bay, with four men I started on foot to the Alcea bay, taking with me a Yacona Indian as a guide. We travelled three miles through the low sand-hills near the southern entrance-point of the Yacona bay. Emerging from the hills, we came upon a hard white sea-beach. The walking here was excellent. Above us rose a wall of high sandstone bluffs, covered with lofty firs and pines, while the ever-succeeding ocean waves rolled and spent themselves at our very feet. We saw and killed several sea birds and bald-headed eagles. We also saw some seals, but did not succeed in killing any of them. We crossed numerous small streams of water, and were occasionally obliged to climb over rocky points extending out into the ocean. The general line of the coast here is nearly south. Leaving the sea-beach, half an hour's walk through some loose sand-hills brought us to the shores of Alcea bay, which is fifteen miles distant from the Yacona.

We built our fire and slept near two Indian lodges, whose inmates scarce knew what to make of our unexpected visit. They appeared to be rather poorer in worldly goods than the Indians of the Yacona, none of the women wearing other clothing than a grass mat fastened round the waist, and some of the men being entirely naked. They had also fewer guns, canoes, &c. There are about thirty of them, in all, living on this river and bay. They say it is five days' hard travel along the coast from this river to the Umpqua. They represent the route as being exceedingly difficult even for men on foot, and as totally impassable with horses, the path frequently climbing the faces of steep cliffs, and passing through the most dense forests. These Indians occasionally visit the Umpquas, with whom they are at peace, for the purpose of buying from them their prisoners, of whom they make slaves. Quite a traffic is thus carried on, the Alcea and Yacona Indians in turn selling these slaves at advanced prices to the Indians living about the Columbia.

September 1.—I went down to the outlet of the Alcea bay in a small canoe, paddled by two Indians. It is only about 80 yards wide, and one-fourth of a mile in length. The tide was falling, and the current setting out so strong that it required the greatest exertion to prevent our little craft from being carried out to sea. The depth of water in the channel was from five and a half to six fathoms. On the north side of the outlet a narrow cape of shifting sand-hills separates the waters of the bay from those of the ocean. On the south side is a sandstone bluff forty feet high. Leaving the beach with a semi-circular sweep, at the distance of about a mile above and below, a chain of lofty breakers stretches completely across the outlet of the Alcea, which would, I think, render it impossible for any vessel to enter the bay. We saw it, too, under favorable circumstances, the sea being generally calm, and no wind stirring.

The bay is from a half to one mile in width. It is bordered by low hills, timbered down to the water's edge. A large part of the bay is left dry at low tide: the average depth of water in the channel, which is narrow and very crooked, is about nine feet. I ascended a few miles up the Alcea, which is shut in by high hills and lofty forests. There are no trails around the bay or up the river. I was informed by some Cliketat Indians that they had once attempted to cut a trail from the Willamette valley down the Alcea river, and had descended within about thirty miles of the ocean, when the country became so broken that they were obliged to abandon the attempt. There are some small fern-covered prairies on the upper part of the Alcea. Returning down the river, we stopped at some Indian lodges, where they had a great abundance of a very excellent little fish, somewhat resembling the sardine in appearance, but larger, which is found also in many of the rivers on the northwest part of this coast, and known by the name of the olhuacan. They take them here in weirs, with large scoops. I saw no indication of coal anywhere on the Alcea, nor any other matter of sufficient interest to require further delay. I therefore pushed back as rapidly as possible, for we had barely subsistence enough in the camp to carry us to the Willamette settlements by the route which I desired to follow, the delays and obstructions to our travel having proved much greater than I had been led to anticipate. But, in fact, our chief hindrance was from the miserable condition of our horses. Had they been in better plight, the trip could have been very easily performed in a third part of the time which it actually consumed. Retracing our steps, we reached the Yacona late in the night, where we found the chief in waiting with his canoe to convey us to our camp on the opposite shore.

September 2.—The heavy smoky fog, in which we had been enveloped since leaving the Willamette valley, partially clearing away to-day, I attempted to make some examination of the outer part of the entrance to Yacona bay. The northern cape of the bay lies further to the west than the southern, and from it there projects out into the ocean a point of low rocks in a south-southwesterly course for the distance of a quarter of a mile: beyond, in nearly the same course, there extends a line of breakers to within a third of a mile of the south shore. The channel runs near the shore on the south side of the entrance. The outer passage is about a mile long, and little over a quarter of a mile in width, bordered on the one hand by a chain of breakers from fifteen to twenty feet high; on the other, by heavy rollers and a low sandy beach. I sounded down this passage not quite half way, carrying from six to seven and a half fathoms of water, when the wind, which was blowing from the northwest, increased to a perfect gale, and, the current setting out very strong, it became too hazardous to venture further. As it was, our canoe got into the edge of the breakers and partially filled with water. When it is calm, however, the Indians frequently go out to sea by this passage; and I think it possible that, under favorable circumstances, vessels could enter the bay. There is, no doubt, sufficient depth of water in the channel of the outer passage, if it be not too narrow and too much exposed. Should it be satisfactorily ascertained that ships may come in with safety, this harbor will become exceedingly valuable, as it is surrounded by a country covered with forests of the finest kind of timber, has good mill-seats, and roads could be constructed which would afford a near market for the produce of the upper Willamette.

September 3.—Moving camp, we came two miles along the shore of the bay; thence striking north, travelled three miles through an open rolling country covered with fine grass and some small patches of fern and thistles. The soil here appeared to be very rich, and was well-watered by numerous little springs. Descending some sandstone bluffs, we followed several miles along the sea beach, until a high rocky point projecting half a mile into the ocean interrupted further travel. We were then obliged to climb along the steep sides of a densely-timbered mountain, at whose base were high perpendicular precipices of volcanic rock, against which the ocean waves roared and lashed themselves with ceaseless fury. Our road was exceedingly bad: in addition to its steepness, immense trunks of fallen trees constantly obstructed our path. It took us over five hours to make about four miles. Notwithstanding the good pasture and rest of several days which the horses had had, three of them utterly gave out and were left behind. Another horse was literally embowelled in attempting to jump a huge fallen tree. We encamped on a small stream in a deep rocky ravine, about 400 yards from the ocean, having travelled 15 miles. I had engaged a Yacona Indian to act as my guide to the Celeetz bay; but, not being closely watched, with characteristic want of faith he slipped out of camp, and I saw no more of him: we were therefore left to find our way as we best could.

September 4.—Soon after leaving camp one of the pack-horses, losing his foothold, rolled two hundred feet down a steep hill, thence over a precipice forty feet high, falling on the solid rocky bed of a small stream which ran below. Much to our surprise he was found quietly eating grass, apparently not being in the least degree hurt, and soon made a second ascent with better success. His saddle, however, and the pack, which contained mess kettles and pans, had not fared so well.

Our road gradually improved as the mountain, receding, left a bench of open land extending from the top of the precipices bordering the ocean to the foot of the steep timbered acclivities, a space varying from onefourth to half a mile in width, well watered, with rich soil, bearing a luxuriant crop of clover, grass, and their usual concomitant the fern. After a few miles travel we again descended to the seabeach, which we followed till late in the afternoon, when, taking a faint trail leading through some low sand-hills, we came to the upper part of the Celeetz bay, where we encamped on a small prairie covered with fine bunchgrass and clover.

September 5.—The day was disagreeably cold, with a dense fog. We attempted to pass round the upper part of the bay, which is bordered by low hills clothed with a dense forest of white and yellow fir, hemlock, spruce, &c. This portion of the bay is a vast marsh, intersected by numerous small canals, which are all filled at high water and left nearly dry as the tide recedes. With considerable difficulty we skirted along the edge of the timber, the ground being wet and lax, and our horses frequently miring down. We at length reached a stream ninety feet wide and from fifteen to twenty feet deep, its margin lined with high bulrushes. I supposed this to be the main Celeetz river. Swimming our horses across it, we made a raft of some dead trees that lay near the bank, upon which we crossed the baggage. Winding our way among the narrow, deep ditches or canals which everywhere obstructed our course, about half a mile further we came to another stream, larger than the one which we had just traversed. The fog having cleared away somewhat, by standing on the backs of our horses we could see yet another large stream beyond, and, as far as the eye could reach, one extended marsh. I saw, therefore, that it would be entirely impracticable to pass round the upper part of the bay, and determined to retrace my steps and endeavor to find some suitable point for crossing lower down. It appears that the Celeetz, at its entrance into the bay, forms a delta, of which we had only passed one arm. In the mean time the tide having fallen, left bare a broad strip of soft mud on each side of the stream already crossed, through which it was impossible to get our horses; we had, therefore, no resource but to wait patiently the rise of the tide. We lay down to sleep on its bank, wet, tired, and disgusted withal at the worse than useless result of our hard day's labor.

September 6.—Recrossing the horses, we extricated ourselves from this marsh and travelled down the shore of the bay. It is about three and a half miles long—greatest width one mile. The opposite shore was almost concealed from view by the fog, but it seemed to be heavily timbered. On the west side it is separated from the ocean by a range of loose sand-hills. It is a custom of the Indians of this country to deposite their dead in canoes, and there are great numbers of them along the borders of the bay. They rest on platforms, each one surrounded by poles, from which are suspended all the personal effects of the deceased.

A chain of lofty breakers extends from shore to shore directly across the outlet of the Celeetz bay, which I think it would be impossible for any kind of vessel or boat to pass in safety. The outlet is only about three hundred and fifty yards wide, and I determined to cross our horses here. Starting them just as the tide commenced ebbing, they were carried by the current, which is very rapid and strong, some six hundred yards down towards a point on the opposite shore, where they all landed safely except one, which, weaker and less able to battle with the waves than its fellows, was swept out into the breakers and immediately drowned. We soon constructed a small raft for ourselves and baggage, the shore being strewn with thousands of drift-logs. It proved, however, so difficult of management, and such a dangerous mode of conveyance in this lightning current, that we were glad to substitute in its stead a fine large canoe which we found concealed among the bushes on the opposite bank. It was after night before all had crossed, and we camped a hundred yards from the shore, at the edge of a pretty grassy prairie which here borders the bay.

September 7.—Early this morning an old Indian entered our camp. He had come in a canoe from some distance up the bay, his attention having been attracted by a large fire which we had built last evening on the southern point of the outlet. He said that himself and another man, with their families, were the only residents on this bay—the last lingering remnants of a large population which once dwelt upon these waters. The mortality of 1831, which proved so fatal to the Indians of the north-west coast, it appears extended its ravages this far south. He told us that we had crossed the bay at the most favorable place, and that it was impossible to pass round the eastern border of the bay with horses. Another large stream coming from the north empties into the bay about half way down its east side; like the Celeetz, it forms a large marsh near its mouth. There are also many other marshy inlets, all impassable and bordered by dense forests.

Taking this Indian as a guide, we travelled round the north point of the outlet and along the sea beach beneath high sandstone bluffs, the distance of two miles and a half; then bidding our final adieu to the ocean, we struck northeast, following a small trail which led us over rolling hills covered with grass and a high growth of fern. About a mile to our right lay a handsome little fresh-water lake, and beyond rose a succession of ridges and tall forests. Having come three miles through the hills, we descended into a fine bottom lying along the banks of a stream about fifty feet wide, to which the Americans have given the name of Rock creek. The soil of this bottom is a dark rich loam. There are no Indians living here.

A large trail crossing the mountains from the Willamette valley descends this creek to the ocean, thence following up the coast to the Columbia. Cattle have been frequently driven by this route from the Willamette settlements to the Clatsop plains, near Astoria. Taking the trail, we ascended Rock creek ten miles, passing over undulating hills and through some thick forests, camping in a small bottom on the north bank of the creek.

September 8.—We continued the ascent of the mountain, travelling through heavy timber. The forest here, as elsewhere in the coast range, is composed principally of the red, white, and yellow fir, different species of pine, maple, ash, yew, and alder. Among the undergrowth there is quite an abundance of the currant, raspberry, blackberry, and serviceberry. Our path was much impeded by logs, brush, numerous rivulets, and mud-holes. Near the Couteau, or summit line of the range, there are many open spots, all covered with luxuriant crops of fern. Descending into the valley of the Willamette, we camped on a fork of the Yam Hill river, at the farm of an American settler, having made twenty-five miles to-day. We were much struck by the contrast in the appearance of vegetation on this side of the mountain, parched and withered by the long drought, while on the west slope we had left it fresh and green as in the early spring. This is, of course, owing to the greater humidity of the atmosphere near the coast. The mountains are by no means so rugged and broken here as where we had crossed them before, and I think it quite practicable to construct a wagon-road to the coast from this point.

Travelling down the Willamette valley, we reached Oregon City on the 13th of September, and on the morning of the 15th I had the honor to report to you in person at Fort Vancouver.

I am, sir, very respectfully, your obedient servant,

THEO. TALBOT,
First Lieutenant First Artillery.

To Bvt. Major J. S. HATHAWAY,
First Regt. U. S. Artillery, Fort Vancouver, Oregon.

Professor John F. Frazer's report on minerals, forwarded by General Smith.

PHILADELPHIA, *March* 21, 1850.

SIR: A few days ago I received from General Persifor F. Smith, commanding the United States forces in Oregon and California, specimens of coal and limestone, with a request that I would examine them and communicate the results to the department. In consequence of which request, I have the honor to submit to you the following analysis:

Specimen 1—*label,* "From Celeetz river, about ten miles from the sea. The vein is under the substance to which a piece is attached, and just below the surface of the water. The river enters the sea about thirty or forty miles north of the Columbia river, found by Lieutenant (?) Talbot, 1st artillery."

Description —Color, black; lustre, resinous; fracture, conchoidal;

splits into rhomboidal fragments; streak, brownish-black, sparkling; density 1.314.

Cokes very easily, and is very dry. The ash is brownish-red and orange, with but little grit.

Analysis.

Loss of weight, at 212°, (water, &c.) - - - -	4.9
Do at red heat, (bitumen, &c.) - - -	49.5
Do by combustion, (free carbon, &c.) -	42.9
Residuum, (ashes) - - - - - - -	2.7
	100

Remarks.—From its physical character and analysis, it is evidently a lignite. The substance to which a piece is attached, as referred to in the label, is a sandstone. The probability of this material being found in large quantity would not be very great, were it not that it occurs so near the deposites of Vancouver's island, which are represented as being so plentiful as to give rise to the hope that it may be a part of the same deposite.

General Smith has given me no further information than that which I herewith submit. If he has reported anything to the department as to the extent or thickness of the seam or deposite, I should be very much interested in receiving the information.

The specimen of limestone is from Monte Diabolo, opposite San Francisco.

Description.—Color drab, and pearl color; fracture, sharp; texture, uncrystalline, but with sparry bands; streak white; density, 2.63.

Analysis by Mr. A. Muckle, my assistant.

Silica - - - - - -	0.3
Oxide of iron and alumine - - - -	1.4
Carbonate of lime - - - -	97.8
Carbonate of magnesia, a trace.	
Water - - - - -	0.25
	99.75

This, then, is a very rich limestone, and in the locality marked for it must prove very valuable to the settler, both for building purposes and for the reduction of the ores of mercury and iron.

It may perhaps interest the department to know that I have received from another part of the country, and further in the interior, a specimen of limestone of very different character.

I remain, with very great respect,

JOHN F. FRAZER.

To the Hon. the SECRETARY OF WAR.

P. S.—Should the department have any further need of my services in matters relating to this subject, please address

JOHN F. FRAZER,
Professor of Natural Philosophy and Chemistry,
University of Pennsylvania.

REPORT OF GENERAL BENNET RILEY.

HEADQUARTERS TENTH MILITARY DEPARTMENT,
Monterey, California, January 1, 1850.

COLONEL: My attention was directed early in the summer of last year
to the importance of a land communication with the Atlantic coast, more
direct and less liable to interruption than any of the routes now used.
The route originally used by the Spanish padres passing through the
Navajo country, but abandoned on account of the hostilities of the Indians, for the Spanish trail, as indicated in Fremont's map, was first suggested to me by Dr. Marsh, for many years a resident of this country.
His representations were subsequently confirmed by others, not likely to
give erroneous information, who represented the route as one by which the
waters of the bay of San Francisco could be reached from New Mexico
"without crossing a mountain." The route was also represented as passing
through a much more fertile country, and by which the desert west of the
Colorado river could be crossed at a point where it is much narrower than
it is on the Gila route.   I considered the subject of so much importance,
that Lieutenant Ord, 3d artillery, whose attention had been directed to
the same subject, was desired to imbody any information that might be
deemed useful.   His report is herewith enclosed, and I respectfully in-
vite the attention of the commander-in-chief to it, in the hope that it may
be deemed of sufficient importance to authorize an expedition for the purpose of determining the practicability of this route for a military or railroad communication.

I entertain but little doubt of its entire practicability, and believe that
it promises advantages very superior to either of the routes now used;
and I urgently recommend that it be explored at the earliest practicable
period.   Lieut. Ord, who goes home on leave of absence, will hand you
this communication, and will furnish you with information in reference to
this subject more in detail, which, I have no doubt, will satisfy the commander-in-chief of the importance of the examination.   It is unnecessary
for me to dilate upon the importance of the route promising the advantages
over any other which this is represented to possess, and of which I entertain no doubt.   An expedition for the purpose of making this examination should be fitted out in the western States, where the animals and
supplies can be obtained at one-third or one-fourth the cost of obtaining
them here; nor would the military force in the department justify the detachment of a sufficient escort, which should be of cavalry, as dragoon
horses cannot be obtained in this country.   I respectfully recommend, if it
should be determined to make this examination, that any recruits of that
arm intended for this department be sent overland, and that they constitute
the escort of the expedition; which might be organized under the command of Capt. A. J. Smith, 1st dragoons, who is in every respect qualified for the performance of this duty, and would, I am assured, be very
much gratified if placed in command of the escort.

For the topographical examinations Lieutenant Ord is respectfully recommended: He is peculiarly fitted for an undertaking of this character;
combining, as it necessarily will, some degree of danger with a great

deal of hardship. His investigations of the subject have put him in possession of a great deal of information that will be highly useful; and knowledge of the country and the Indians on this side of the Colorado, where the greatest difficulties are to be apprehended, would tend greatly to insure its success. In addition to these reasons, Lieutenant Ord has been mainly instrumental in collecting the information upon which this report is based, and I think it very proper that any credit that may result from the successful prosecution of the expedition should accrue to him.

Very respectfully, Colonel, your obedient servant,

B. RILEY,
Bvt. Brig. Gen. U. S. A., Comm'g the Department.

Lieutenant E. O. C. Ord's first report to General Riley.

MONTEREY, CALIFORNIA, *October* 31, 1849.

SIR: In obedience to orders, I left this place in the steamer of the first instant, landed at San Diego on the third, and next afternoon (as soon as horses could be got) proceeded over some seventy-five miles of clay hills, by the shortest road, to the plains of Los Angeles. These plains have an average width of about thirty miles; length about seventy-five; watered by three short rivers running from the mountains to the sea in the winter, and sinking in the plains in summer. Here are the principal cattle ranchos of the lower country, into which the Indians enter to steal, first from the Pa-Utah country, on the east by the Cajou; second, from the Tulare or San Joaquin country, by the San Fernando pass; and, third, from the Corvilla country, on the southeast, by the San Gorgona and Warner's passes. As the Pa-Utahs are the greatest thieves, and most cattle are driven out by the "Cajou de los Mexicanos," I turned up the Santa Anna, and proceeded at once to the rancho of Colonel Williams, examined its houses and grounds, and next day crossed the plain between this rancho and the Cajou to "Jurupa," on the Santa Anna, twenty-one miles; thence I followed up the river to Lapolitana, via the Mexican Pueblo, seven miles; thence up Cajou creek to pass, ten miles; thence two or three miles in the pass. At the mouth or entrance of the pass, and for a mile or two on either side, the land, though flat, is stony and barren, and four or five miles out in the plain without grass, and covered with "chemisal" or low bushes. There is water in the pass, but not grass to support a picket of cavalry; and even should one be posted there, it would not prevent the entrance of the Indians, who could easily avoid such picket by a detour of about a league to the right, by another though not very good pass, called "El Cajou de los Negros." After examining the pass, I attempted to cross the plain in a direct line through the chemisal to Williams's; but night came on, my guide got into some hills on the left, got lost, and I had to lead him to Bandini's rancho, on the Santa Anna. Next morning returned to Williams's, found the buildings ample for one company, the position facing the Cajou, eighteen miles off; no intervening hills, and no time to put up quarters on the Indian side of the pass. I contracted for quarters there, (see contract,) after which I rode some eighteen miles to reported coal beds; found it bitumen, hardened. Next day I returned to Bernardo Yerba's, the largest and most reliable farmer in the neighborhood; contracted for three hundred fenegas of corn, (all he could spare.) I found no one else in the vicinity willing to make large contracts, and I deemed it best that small purchases should be left to the quartermaster. From Yerba's I pushed on to the Pueblo, making inquiries as to the disposition of the people to defend their property in case arms and ammunition should be issued them. They are willing; and, if properly organized, would be able to do so against all Indian incursions, but there is no unity of action among them.

At the Pueblo I spoke to the judge, the alcaldes, Messrs. Wilson, Temple, Pio Pico, Manuel Riquena, and other of the principal inhabitants, urging the formation of a volunteer company, or some effort at organization, that they might avail themselves of arms, &c. to be issued them. They all acknowledged the propriety of such a measure; said they would consider it, and would no doubt form some company. I think a supply of arms, placed in charge of some responsible person, to be issued when required, or upon the organization of their community, an advisable measure under present circumstances. As I was anxious to get back in time to enable such measures to be taken as might be thought proper before the rainy season, I returned to San Diego, waited there for the steamer of the 28th, when, being very impatient, I hired horses and was about starting up by land, when the steamer hove in sight; I got on board, and arrived here this morning.

Allow me to state that, though Williams's rancho appears the best point in the vicinity of the Cajou to quarter troops before the rainy season sets in, I do not deem it the proper place for a permanent garrison to protect that district from the Indian horse-thieves. It is reported to be but two days, with wagons, through this pass to the Pa-Utah country on the Mojave, the valley of which is large, well watered, with good timber for building but four or five miles from the water-course, and abundance of fine grass. It is on the most direct route for New Mexico and the Salt Lake, in the plain, which opens three important and easy passes—that to the north, into the San Joaquin plains; that to the east, leading into the Great Basin and Indian country of the Apaches, Utahs, and Navajoes, and the Cajou to the Pacific. The upper end of the San Joaquin district, where the San Fernando and Tejou passes enter, would require other defences. The San Gorgona and Warner's passes appear to require no permanent establishment. The horrors of the Gila route will prevent its use after this year; and the Indians in that quarter are not yet hostile. I further report that many emigrants arriving at San Diego are destitute; neither bread, flour, nor sugar can be purchased there. I requested the commanding officer to allow sales in small quantities to them until your orders in the case could be received.

The sketch to the above report will be handed in as soon as completed.
Very respectfully, &c.,

E. O. C. ORD.

To Major E. R. S. Canby,
  Asst. Adjt. Gen., 10th Military Dept., Monterey, California.

Lieutenant E. O. C. Ord's second report to General Riley.

MONTEREY, CALIFORNIA, *December* 30, 1849.

SIR: In answer to your request, that I should condense such information as I may have collected upon the facility of constructing a rail or other road through the country lying between the southeast corner of California and the valley of the Upper Rio Grande, allow me to state: The sad condition in which immigrants arrive by both the land routes to California led me to inquire if a better intermediate route did not exist, over which I knew the padres, during the Spanish times, used to pass.

The Mojave valley is a sort of continuation of the Tulare, is the nearest habitable and fertile district of California to New Mexico, and is connected, by the passes of Walker, the Tejou, and "Cajou de los Mejicanos," with the most fertile of the great plains of the Pacific. Though this country is thickly populated for an Indian country, the Indians are so peaceable that parties of two, three, and four whites are in the continual habit of travelling therein; and it was from this district that (I was invariably told by the oldest and best informed trappers) led the shortest and best road to New Mexico, entering the valley of the Rio Grande somewhere opposite the town of Abuquerque. To connect these I have resorted to all the oldest maps and mountain men within my reach; and I have learned that, leaving the upper end of the great San Joaquin or Tulare plain, via Walker's pass, parties have gone through a well watered, timbered, and habitable district, without crossing any mountains, to the Mojave, or Rio de los Martires, (which rivers I think are one, though some maps separate them,) about one hundred and forty miles. Thence following down the course of this river, (which runs through a fertile valley, with abundance of timber on its banks,) about sixty miles ; thence striking across the sand-hills and plains, about ninety or a hundred miles, to the "Paso de los Padres" of the Colorado, which is represented as a good pass or ford, lying at the head of "Long Cañon;" thence taking one of two valleys—one of which comes in from the northeast of the "Paso," and runs at about five leagues distance from the river, and nearly parallel thereto, and the other comes into the valley of the river about ten leagues to the east of the Paso, and runs also northeast and southwest, at about ten leagues distance from the river. By the first of these it is about forty, and by the second fifty leagues to the valley of a small tributary of the Colorado; and, following up this valley, heading the stream to the north, striking the valley of the Jaquesita at or near the head of the first named stream—following up this valley, it is about thirty leagues (as above) to Oraybe; thence are pueblos, at short intervals, in the Moqui country; and a road, which I find laid down in an old English map, via Osoli, Casita, to Tuñi, which is on the eastern slope of the Mimbres, and has a gentle, practicable (to R. road) descent to Acoma, in the valley of the Del Norte. The comparative freedom of this route from deserts and mountain ridges depends upon the statements of all the best informed and oldest trappers with whom I have conversed. The country was partly passed over by Padre Garces in 1775, whose course, as far as Oraybe, I have described from a manuscript map obtained from the former secretary of state in this country, and which is the original of Padre Fout, dated 1777. The piece of sand-hills lying between the Mojave and El Paso de los Padres can, I am informed by an old trapper, be partly avoided by leaving the Mojave on the present well known Mexican trail, following that for about three days, and then striking off to the east-southeast. The only difficulty to opening this road at once results in the large and numerous tribes of hostiles between the Del Norte and the Colorado; and this, if properly viewed, is a fortunate and inevitable difficulty, to be sought and encountered as one of the greatest recommendations in favor of the country in which it exists. A population is a proof of the fertility of the district, and is found on none of the other routes to this country, for such roads were opened by solitary trappers or small parties of Mexicans who sought deserts rather than encounter Indians. If the people of the United States want a road from New Mexico to this country, they would not select a hot, cold, or sterile district in which to build it, or be deterred in the work by the knowledge that the best country for it is occupied by powerful or hostile Indians. Besides, the very means necessary to pacify such hostiles are the most certain to lead to the opening of a good road, the rapid rise of towns and villages, and the final completion of such means of communication through the land as the age, the wants, and abilities of the people might suggest—I mean the establishment of a line of forts through the country as soon as an official examination thereof might authorize it.

Respectfully, your obedient servant,

E. O. C. ORD,
First Lieutenant 3d Artillery.

Major E. R. S. CANBY,
    Assistant Adjutant General,
        Tenth Military Department, Monterey, California.

MONTEREY, CALIFORNIA,
November 6, 1849.

SIR: In compliance with part of my instructions received from department headquarters, in connexion with my reconnoissance of the Cajou pass, I have the honor to report that I landed at San Diego. Here there is a good frigate harbor—the quietest on this coast. No good survey has yet been made of it by the English; one such is very much wanted, as vessels from China and from the Islands make this part of the coast first, and have to beat up to San Francisco against the prevalent northers. The steamers have established their coal depot at this place in preference to San Francisco. The entrance, about two hundred yards wide, is of easy defence. The town, containing about twenty badly built adobe houses and two good ones, is five miles from the anchorage; a road leads around the bay shore, bordered by naked hills. Small boats cannot get within half a mile of town, owing to shallow water and marsh mud. The town was built here because it is the only point on the bay where a good supply of water can be had from the little river which here in summer empties its waters into the bay. In the autumn these waters dry or sink into the sandy bed. I followed this bed of sand through its valley amongst low cactus, covered clay hills, seven miles, to the mission. There I found the old walls tumbling in, and everything going to ruins. The vineyard and olive trees still produced a few fruit. The library should be placed in charge of some one interested in its preservation. From the mission along the river bed it is about five miles more to a gap in the hills, where the valley heads. Near this gap are a few sycamore trees and willow bushes, the only timber in the valley. This valley, though much smaller than that of the Carmel, has a good soil, and if cultivated would perhaps maintain three or four hundred inhabitants. There is still in good preservation an aqueduct leading from the running water in the hills to the mission, which with a little care would supply water for irrigation to the whole valley. The rocks at the gap above the mission appear to be of trap recently upheaved. Between this point and the beach, the gravel and clay hills are composed of a sedimentary deposite. Coal was reported to exist near San Diego, but the late Captain Warner upon examination found it to consist of a small deposite of lignite.

The bay shore south of San Diego is bordered by the same low gravel and clay hills already described. No timber here, but little grass; plenty of cacti, occupied by hares, rabbits, and quails.

From San Diego I took the road northwest up the coast towards Los Angeles, which was passed around the edge of False bay, a shallow arm of the sea about three or four miles north of San Diego; no hills between them, though a bluff separates the two hills near the sea. From the False bay the road leads for thirty miles amongst the low table hills of the coast, similar in formation to those above described, and equally as barren, though at intervals of from four to ten miles it crosses little valleys, which when watered contain an adobe hut or two, some small patches of corn, beans, and pumpkins, and a few cattle. The valley of San Diegito, twenty miles from San Diego, if all cultivated, would produce several thousand fanegas of maize, and other things in proportion. From the half dozen patches I saw there, the Indians were gathering from twenty to forty fanegas of maize to the acre—this without irrigation. About thirty-six or forty miles from San Diego, in a similar valley near the sea-side, the road passes the now deserted mission of San Luis Rey. It was well built, and its handsome church, extensive arcades, orchard and vineyard, still look fresh: the people are all dead or gone away. I found that the birds and the sunshine had entered by a hole in the roof; and one little gorion was perched on the altar, singing away as if the place belonged to him. Some one interested in their preservation should be sent there to protect the wood work and gardens from the ravages of travellers. Three or four miles to the north of San Luis is another and better watered valley, occupied by a few Indians and Don Pio Pico's rancho. This gentleman has a pretty good vineyard, though he cultivates but little.

The valley would maintain some thousand souls. Near this valley, where the road crosses a spur of rocky coast hills, I observed what I took to be marl or rotten limestone. Five miles from the valley the road again strikes the beach at the little mission of Las Flores—as usual, going to ruins; thence it follows the beach for twenty or twenty-five miles to the valley of San Juan Capistrano, keeping aforesaid coast range close on the right. These rocky hills appear to be of the same formation as the volcanic ridges in the Stanislaus gold region. I observed a prevalence of talcose slate, interspersed with quartz veins and boulders, supersoil of red clay and igneous gravel, all of which are characteristics of the gold soil. Gold is said to have been found here.

In the pretty little valley of San Juan are the remains of the mission and church of that name. The latter was built with an effort at ornament, as the handsome cut-stone cornices, pilasters, and medallions still show. During the great earthquake of 1814, one-half this building fell, and crushed to death a number of persons in its fall. There are still eighteen or twenty poor families living in the valley. They raise an abundance of corn, pulse, &c., and some very good peaches.

From San Juan the road turns inland and leads northward for six or eight miles, through low prairie hills, to the south end of Los Angeles plains. These plains extend from San Juan, on the southeast, to Canenga, on the northwest, a distance of about seventy miles; on the left the coast and the harbor of San Pedro, (no survey of them yet made except by the English;) on the right a range of low hills; and further inland the face of the Sierra Madre and San Bernardino, from which flow into the plains three small rivers—the Santa Anna, the San Gabriel, and the Los Angeles. Their waters last with but little intermission; though in the fall they come only a short way into the plains, when they sink in the sands, to rise again lower down the plain in some dry springs and wet places.

In the spring these plains are covered with grass, geese, flowers, and cattle. The grass dries in the summer, the geese fly away, the flowers die, and the cattle remain to fatten on the hay. Formerly, they used to kill the cattle in the fall for their hides and tallow, but now there are no more hide droguers, and the cattle have to perform long journeys north, to carry beef to the mines.

I skirted the plain near the hills on the right for about fifteen miles to where the Santa Anna came in—passing one rancho, a cattle farm. Shortly after entering the plain, the waters of the Santa Anna sink in its sandy bed. Here the road continues on to Los Angeles, but I turned up the sandy bed of the stream; and in about half an hour's ride, the valley and its grove of trees opened before us: at the mouth of which valley, on the rancho of Theodosio Yerba, are three or four snug houses, a good vineyard and orchard, and about ten or twelve acres under cultivation, (he might cultivate ten or twelve hundred,) from which he expected three or four hundred fanegas of corn. Here I found the river a very respectable stream, crossed it where it had a rapid current, width of sixty to a hundred yards, and a depth of from six to twenty inches of sweet, clear water. It is the largest river south of San Francisco bay, in California; and though I had been nearly three years in the country, I did not think there was such a pretty valley in it as that which I now entered. It is from one-half to two miles wide, with low, verdant banks, covered with groves of cotton-wood, willow, and sycamore. An hour's gallop carried me to the house of Don Bernardo Yerba—large, clean, and comfortable. He has a large vineyard, and thrice as much land under cultivation as his brother Theodosio; could not say how much corn he would gather, as he did not expect to get it in before November: it was then the 9th of October. The people here frequently plant maize in July and August, when they are eating roasting-ears from the first crop. In this and the neighboring valleys, owing to the abundance of water, and these waters being nearly even with the river bottoms through which they flow, irrigation is but a trifling labor, and the farmer need not fear losing his crop by too much rain. From Bernardo Yerba's I took a short cut across the hills to the Chino rancho, eight miles off. From the top of these hills I had a fine view, to the rear, of the plains, the coast, and the islands abreast. In front, at my feet, lay Williams's buildings and rancho, at the edge of a bright green little plain. Eighteen or twenty miles across this plain rose the lofty Sierra. It looked a half an hour off, and took me all day to ride to it. The gap called the Cajou pass, or the Cajou de los Mejicanos, was distinctly visible; course from the hill-top, northeast. To the left of this gap about a league was another, also leading into the Pinte's country, called the Cajou de los Negros. To the right of the main or Mexican Cajou rose the San Bernardino, the only mountain to the east of the Sierra Nevada, on which the snow lies till August, and which can be seen at Los Angeles, sixty miles off. Still to the east of San Bernardino, my guide pointed out the San Gorgona Pass, leading into the Corvilla country and on towards Red river. The range to the west of the Cajou de los Prietos, or Negros, is continuous, like a great wall, for about fifty or sixty miles, to the pass of San Fernando, or "Cajou de los Uvas:" this leads into the head of the Tulare or San Joaquin plains. To the east of the Cajou de los Negros the range is broken into isolated peaks, with intervening valleys. To the southeast was seen the emigrant's road, coming in from the Gila and the desert. Directly through the Cajou led northeasterly the way to the Great Basin and the Salt Lake. After taking the bearing of all these, I descended to Colonel Williams's rancho and a good breakfast. The Colonel has a large and pretty place, which, being on the lower or damper side of the valley, preserved its greeness late in the fall. He has a good orchard and vineyard; and I have forgotten how many thousand fanegas of wheat he said he could raise without irrigation. This place being the crossing point of two important routes, it may be the site of some future town. I rode across the south side of the valley, striking the river again at Bandini's rancho; thence I skirted along the river bank two or three miles, when I struck another spur of hills which abutted on the river. Crossing these, I descended into the river bottom at a place called Jurupa, a flourishing rancho about twenty miles from Williams's; thence following the river, still bordered with groves of cotton-wood, sycamore, walnut, and willow, I passed a settlement of New Mexicans a league above Jurupa, and two more leagues brought me to where the river forks—the left hand fork called Cajou creek. On this is a flourishing settlement of Corvilla Indians—about forty lodges. They were getting in their corn; and I was pleased to see that their industry had been rewarded, for they were housing about six hundred fenegas of fine corn, and plenty of large pumpkins.

I followed Cajou creek northwestward about two leagues, until arriving opposite the Cajou, (where the land becomes very stony and sterile, covered with the debris washed from the mountain gorges;) here I turned into the Cajou, and made the examination already reported. On my return, my guide, in attempting to cross the plain, got benighted, and took to the hills on the southeast of the valley, the spur of which I had crossed near Jurupa. I noticed, in galloping through these hills, large boulders of well-worn granite rolled from their beds, and well-worn corners of the same projecting from the earth, (evidently of an antediluvian birth;) also boulders and veins of hard limestone. Next morning I got back to Williams's, and spent the day scouring the hills between his place and the coast, with one of his assistants, after the coal which he said had been lately discovered. We could not find the coal, but got stuck in several pitch springs, which article abounds in this formation all along the coast. These hills are of red sandstone, and overlying clay-slate, dip about 70° seaward; both strata much shattered by earthquakes. Next morning I started for Los Angeles, via Santa Anna, the road from which river led across the plain, near the hills, (passing three or four grazing farms,) to the river San Gabriel, about 25 miles. The waters of this stream run a few miles in the plain before sinking. It is similar to the Santa Anna, but with a shorter course, less water, and the banks in the plain better land. On this river are the remains of the old mission of the same name, formerly one of the richest in California. There are still cultivated some fields of corn, and a few vineyards; though the farmer's wealth here, as elsewhere in California, consists in his stock, which requires no labor. From the San Gabriel it is about eight miles further to the town of Los Angeles, where the river of the same name comes out into the plains. For about four miles up and down the stream the river bottom is covered with vineyards, orchards, and gardens, through which wind lanes leading to town. This latter is built close under the hills on the north side of the valley, and consists of an old adobe church, and about a hundred adobe houses scattered around a dusty plaza, and along three or four broad streets leading thereto.

This place, until the discovery of the placer, by its tropical climate, abundant waters, and fertile plains, had attracted a population from the northern provinces of Mexico of about three thousand, among whom were a few retired trappers and a number of vagabond Indians, the refuse from the missions. To give an idea of this as well as other parts of the California population, I may as well sketch the rise and fall of the above-named missions.

This whole coast, a little less than a century ago, was a howling wilderness, covered with Indians and animals. Spanish missionaries came, gathered the Indians and tamed them, set them to building big churches and snug huts, made them cultivate the soil and raise great herds of cattle. Their lands, extending from mission to mission, covered all the coast from the sea back to the mountains, and each mission had from two to six thousand proselytes. To take care of these,-they sent for Spanish soldiers. A corporal's guard for each mission came. Unluckily, they brought wives and daughters. Everything in the country bred well; and, before the Mexican revolution was felt here, a mongrel population of rosy-cheeked, buxom girls, hard-riding, hairy-headed boys, began to grow up. The fathers of these wanted some place for their cattle and families; and, to prevent confusion, the padres assigned them parts of the pretty valleys of San Jose to the north, and Los Angeles to the south. These pueblos, called the Upper and Lower, were increased from abroad, and multiplied at home; and, when they heard of the revolution, delegates were sent to Mexico. Soon the missions were secularized—the Spanish padres turned out, Mexicans turned in. The cattle and what Indians did not run away were divided up amongst the descendants of the corporal's guard. About this time hungry, blue-eyed men, who had wandered from the east after beaver, came over the desert and down the mountains into these pleasant plains, where they found plenty of fat beeves and pretty girls. Unluckily again, these latter preferred the new comers; and the new comers became good Catholics, and fathers of large families. When one of these beaver trappers goes off, and don't come back, others, thinking he has found a good place, follow; so that, since the arrival of the first trapper from the east, tracks have led this way till the new comers and their countrymen (come by sea) outnumbered the descendants of the corporal's guards. But, more polite than the padres, who took the Indians' lands, or than said descendants, who took the padres' lands, they sought unoccupied ground, and have driven out no one whom they found in quiet possession. As nearly all the good lands near Los Angeles were occupied before they began to come in, but few blue-eyed people are found there; and, until these broad plains are settled by a race of people accustomed to the gratification of more wants, they will remain as they are, almost unproductive, in the hands of those satisfied with the gratification of a few. I think the first class of men will buy up these lands, when they shall discover that, attached to these plains, at San Pedro is a good harbor for coasters and steamers; that the sea is quiet in summer, navigable for open boats; and that now a good market is at hand on the plains for all the vegetables produced on them, at ten times the prices paid in any other part of the United States.

I reckoned about ten grazing farms on the Santa Anna, containing, in all, about 40,000 cattle, and 1,000 horses; about the same number on the Los Angeles, and about twice that number on the San Gabriel. There are from sixty to a hundred cornfields and as many vineyards around Los Angeles, producing from three to four thousand fanegas of corn, and as many barrels of wine.

The port of San Pedro is about twenty-five miles off; gentle slope down the plain whole way; very good building-stone in the harbor and on all the hills next the plain.

Whilst examining the country in the vicinity of the Cajou pass, I was only prevented from going through it and examining the entrances of the Great Basin by want of time. I have, however, made many inquiries upon its occupants and resources, and would suggest that an examination be made, by a strong force, of the Indian country between the Mojave and Santa Fe, as soon as possible.

'I am convinced that if a line of forts can be established across this country, it will be the means of opening a short winter and summer route between the United States on the east and California on the west. The country is reported habitable, passable, and with an outlet near Jemes, New Mexico, for wagons. If such an outlet exists, the Gila route, now covered with carcasses and lined with graves, will he abandoned. We will hear of no more cannibalism in the snowy mountains; and Americans in this country will feel and act as if they were still in the old United States.

Very respectfully, your obedient servant,

E. O. C. ORD.

Major E. R. S. CANBY,
Assistant Adjutant General, 10th Military Department.

(PART II)

REPORT

OF

THE SECRETARY OF WAR,

IN FURTHER COMPLIANCE WITH

The resolution of the Senate calling for copies of reports on the geology and topography of California.

JUNE 11, 1850.
Referred to the Committee on Printing.
JUNE 24, 1850.
Ordered to be printed, and that 5,000 additional copies be printed for the use of the Senate.

WAR DEPARTMENT,
Washington, May 17, 1850.

SIR: In further compliance with a resolution of the Senate of the 25th of February, requiring copies of reports on the geology and topography of California, I have the honor to submit, herewith, copies of reports, with accompanying maps, of officers of the corps of topographical engineers, transmitted to this department by General Persifor F. Smith and General Bennet Riley.

A report under the above resolution was made by me on the 27th of March, and I have the honor to request that the enclosed papers may be taken as part of that communication.

Very respectfully, your obedient servant,
GEO. W. CRAWFORD,
Secretary of War.

Hon. WILLIAM R. KING,
President pro tempore of the Senate.

P. S.—*June* 10.—The accompanying papers with this report have been accidentally delayed at the department till this date.

[No. 6.]   HEADQUARTERS TENTH MILITARY DEPARTMENT,
Monterey, California, January 30, 1850.

COLONEL: I have the honor to transmit, herewith, a copy of Lieutenant Derby's map of a portion of the valley of the upper Sacramento.

The memoir accompanying this map will be transmitted as soon as it can be copied.

Very respectfully, Colonel, your obedient servant,

B. RILEY,

Bt. Brig. Gen. U. S. A., commanding the department

Lieut. Col. W. G. FREEMAN,

    Assistant Adjutant General U. S. A,

       Headquarters of the army, New York.

[No. 10.]   HEADQUARTERS TENTH MILITARY DEPARTMENT,

       Monterey, California, February 14, 1850.

COLONEL: I have the honor to transmit, herewith, a copy of the memoir by Lieutenant Derby, topographical engineers, of his survey of a portion of the valley of the Sacramento. The map of this survey was transmitted with my letter (No. 6) of January 30.

Very respectfully, Colonel, your obedient servant,

B. RILEY,

Bt. Brig. Gen. U. S. A., commanding the department.

Lieut. Col. W. G. FREEMAN,

    Assistant Adjutant General U. S. A.,

       Headquarters of the army, New York.

Topographical memoir accompanying maps of the Sacramento valley, &c.—Respectfully submitted by G. H. Derby, lieutenant topographical engineers.

       HEADQUARTERS TENTH MILITARY DEPARTMENT,

         Monterey, California, September 5, 1849.

SIR: Major Kingsbury, 2d infantry, was instructed, from division headquarters, on the 10th of July last, to establish his command at Johnson's rancho, on Bear creek, a tributary of Feather river, and to lay off a reservation of one square mile, &c. You will accordingly, after joining Major Kingsbury, proceed to survey and mark out this reservation; and you are authorized to call upon Major Kingsbury for any assistance that you may require for that purpose.

Two copies of your map will be made—one of which will be left with Major Kingsbury for the post, and the other will be forwarded to the department's headquarters.

The commanding general has learned unofficially that Major Kingsbury has been authorized by General Smith to select any point he may deem preferable to that indicated in the instructions above referred to. In this case you are directed to render Major Kingsbury any assistance that may be in your power, in enabling him to make this selection. You will at the same time make a topographical sketch of the country that you may examine for this purpose, and collect and report, for the consideration of the commanding general, any information that may be useful with regard to the resources of the country, the means of communication, the number of Indians, (distinguishing between the tame Indians of the rancho and the wild Indians of the Sierra,) and the comparative advantages of different positions for military posts, having particular reference to healthy locations, &c., &c.

In reporting upon the resources of the country, you will designate particularly those that may be useful to a military command, such as forage, building materials, &c.

After the completion of these duties with Major Kingsbury's command, you will make an examination of the valley of the Sacramento to about the latitude of 39° 20′, or the mouth of Butte river, collecting and reporting any information that may be useful upon the subjects above referred to. It is not designed at this time that you make a detailed survey of the country you examine, or that your report should be confined to personal observation; on the contrary, the General desires that you will imbody in your report any reliable information that you may derive from the inhabitants of the country, as well in relation to the *general* as to the *military* resources of the country.

It is supposed that you obtained at Benicia the instruments, &c., that you require for the performance of this duty. If, however, you were unable to do this, you are authorized to purchase such as are indispensably necessary. If without funds, you are desired to estimate at your earliest convenience for funds, in order that the commanding general may make the necessary arrangements for supplying you.

You are authorized to employ three assistants at the average wages of the country, and to purchase the horses and pack animals that may be necessary.

So soon as these duties are completed, you will report in person at department headquarters; and the General directs me to impress upon you the importance of completing them at the earliest possible period, as your services are greatly required at other points.

These instructions will be sent in triplicate.

Very respectfully, sir, your obedient servant,

E. R. S. CANBY,
Assistant Adjutant General.

Lieutenant G. H. DERBY,
Topographical Engineers.

———

TOPOGRAPHICAL OFFICE, HEADQ'RS TENTH MILITARY DEP'T,
Monterey, California, December 1, 1849.

MAJOR: In compliance with the above instructions, which I had the honor to receive on the 17th September, on my arrival at camp Anderson, near Sutter, California, I immediately reported to Captain H. Day, who I found had relieved Major Kingsbury, in command of the battalion of the 2d infantry, stationed at that point. I had not received department special orders No. 44, dated September 4; and having, therefore, no previous information of the duties assigned me, was entirely unprepared with either instruments or funds for their execution.

Finding, however, that Captain Day was about to move from camp Anderson, and my services being required in the selection of a site for the new military post contemplated, and having no time to communicate with department headquarters, (as, by *ordinary* means of communication, a month would be required for that purpose, and an express could not be sent at a less expense than $1,000,) I concluded to purchase on credit such instruments, animals, and other indispensables as might be required for the survey, and to forward to the department, at once, my estimates, trusting that they would be received and honored before my return. This course, I may add, was in accordance with the advice of the senior officers on the station, whose opinion I of course solicited immediately upon receiving my instructions.

I therefore proceeded at once to engage the services of an assistant, three rodmen, and a teamster, who, with my servant and a gentleman named Kemp, who volunteered to accompany me as far as Bear creek, composed my party. I had succeeded in procuring a circumferentor and chain left by Captain Warner (before starting on his late melancholy and fatal expedition) in the hands of Judge Schoolcraft, at Sacramento City, but was compelled to purchase a chronometer, sextant, artificial horizon, and an additional compass and chain. I then procured from Messrs. Smith, Bensley, and Company a wagon with six mules, for the transportation of our instruments, subsistence stores, forage, &c., the necessary riding animals for the party, and such articles of mess furniture, &c., as we required; for all of which, together with five hundred dollars, which they advanced me for the expenses of the road, they very kindly gave me credit. I was obliged to pay a much higher price for my mules and wagon, from the fact that the expedition for the relief of the emigrants, under the command of Major Rucker, was at this time being fitted out at Sacramento City, and his large purchases of animals and means of transportation had created a scarcity of both throughout the adjacent country.

We left Sacramento City at 2 p. m. on the 22d of September, intending to make a short march in advance of the infantry command for the purpose of trying our mules, those composing the team having never been worked together in harness.

Leaving Sutter's fort about half a mile to our right, and crossing the bottom-land between the Sacramento and American rivers in a direction N. 30° E. from the city, we arrived at the lower ford of the American at 5 p. m., having made a distance of nearly two miles in three hours, without further accident than breaking the tongue of the wagon, which we repaired upon the spot by wrapping the fractured part tightly with rope. This was owing to the refractory conduct of our team, which, upon arriving at the ford, utterly refused to pull the wagon across; so, after exhausting the usual arguments, manual and vocal, resorted to in such cases, I was fain to hire an individual, who was luckily passing at the time with eight yoke of oxen, to assist us, which he did with such effect that we soon found ourselves on the opposite bank of the stream, with a smooth road before us. The American river, at whose junction with the Sacramento the city is situated, is at this ford about three hundred yards in width; the banks are some thirty feet in height, but gradual and easy of ascent, which is not the case at the ford (Child's) some eight miles above. A sand-bar, which at extreme low-water is exposed, forming a small island in the middle of the river, makes out from the southern bank. At the time of our crossing the water was quite low, varying from eighteen inches to two and a half feet in depth; but at the commencement of the rainy season it swells rapidly—three days of heavy rain being sufficient to raise it from four to six feet. The tide rises and falls at Sacramento

City, causing a variation in the depth between high and low tides of from six to fourteen inches. On crossing the American, we passed through a fine grove of oaks which borders the stream through its entire extent, and, striking a more northerly direction, arrived without further accident at Dry creek, which at this time was, with the exception of a few holes, perfectly dry, but which in the rainy season is quite a considerable stream, rising in the spurs of the Sierra Nevada and running in nearly a westerly direction to the Sacramento. Its distance at our point of crossing from the ford of the American is six miles and a half. Here we encamped for the night, finding water in considerable quantity about one and a half mile east of the road in the bed of the creek. About 9 p. m. Captain Day's command came up and encamped in our vicinity. The plain between the American river and this creek is of dark alluvial soil, and, with the exception of a range of low sand-hills running parallel to, at a distance of six miles from, the Sacramento, is an almost unbroken level, extending east about twenty miles, where it rises into low hills, which commence the western slopes of the great Sierra Nevada. A small field of *tulé* commences with the Dry creek, extending nearly to the Sacramento. Upon the commencement of the rainy season the soil upon this plain greedily absorbs the water, and in a few days becomes a thick, tenacious quagmire, which it is difficult, not to say dangerous, to attempt crossing, even with pack-animals. The *tulé* at this time is preferable for crossing, as its thickly-interlaced roots, until thoroughly saturated with water, continue elastic, affording for some days a safe passage to terra firma.

We made observations for latitude and longitude by meridian altitude of the sun on the 23d, and left our encampment at 5 p. m., proceeding over a plain of precisely the same character as that of yesterday, with no alleviation of its unbroken surface to the eye but the distant outline of the timber bordering the Sacramento and Feather rivers.

After marching fifteen miles N 10° W. (by the compass) we came to a small pool of stagnant water, where we found several emigrant families encamped, although there was no wood or grass in the vicinity; and proceeding one and a half mile further on, we came to a small pond, where we encamped at 10 p. m., managing to procure a scanty supply of wood for our camp fires from the stunted alders which surrounded the muddy and unwholesome-looking water. Leaving our encampment at 8 a. m., we soon arrived at the rancho of Nicholas Altegier, situated in a fertile spot at the junction of Bear creek and Feather river. Mr. Altegier is an old resident of the country, and his farm at this place being a well known position, has from time immemorial been called " Nicholas Rancho," he himself being universally known as " Nicholas" only. He has a field enclosed containing about a quarter square mile, apparently of the most fertile soil, and owns a large adobe house of two stories in height, which presents quite an imposing appearance in this country of log-huts and Indian rancherias. About 100 wretched Indians, playfully termed Christian, live in the vicinity upon the bank of the Feather river, subsisting upon acorns (which, pulverized with roasted grasshoppers, they form into a cake) and salmon, with which delicious fish the river abounds. The more intelligent and docile of these creatures are taken and brought up on the farm, where in time they become excellent *vaqueros*, or herdsmen, and where they are content to remain, receiving in return for their services such food and clothing as it may suit the interest or inclination of its owner to bestow upon them. About one mile south of "Nicholas Rancho" the road divides—the right hand path leading directly across the plain to Johnson's rancho, (now Gillespie's, on Bear creek,) the left passing "Nicholas Rancho," and crossing the creek about one and a half mile from its mouth. A path leads from this crossing directly up the bank of the creek, and joins the right-hand main road again about six miles south of "Gillespie's."

The right-hand main road is the " Truckee route," or emigrant trail, from the Salt Lake via Truckee river; the left is " Lawson's route," or the northern emigrant trail, entering at the head of the Sacramento valley, near the headwaters of the Feather river. This latter route is some 300 miles farther than that by Truckee river, but has the advantage of easier ascents and descents, and generally affording better pasturage for cattle

Leaving "Nicholas Rancho," we continued upon the river road, passing the crossing, and encamped at a distance of ten miles from the house upon Bear creek, in a beautiful grove of oaks and sycamores, surrounded by high grass which borders the creek for some two miles in depth on either side, and which afforded to our tired animals the most extreme satisfaction. Here Captain Day concluded to leave his command encamped, while making an examination of the country in the vicinity; this point being within five miles of the first proposed in our instructions, and within easy riding distance of the remaining points proposed for selection. Accordingly, leaving my company in charge of my assistant, (Mr. J. H. Newete) I accompanied Captain Day on the morning of the 25th, for the purpose of assisting in the selection of the site for the proposed military post. Proceeding (by compass) in the direction N. 45° E., we arrived at Johnson's rancho, a small one-story adobe building, at a distance of five miles from our encampment upon the north bank of Bear creek. Here we were kindly received and entertained by the proprietor (Mr. E. Gillespie,) who volunteered to accompany us in our examination of the rancho. We accordingly rode over and made a general reconnoissance of the country in the immediate vicinity, embracing some ten or twelve square miles, and, returning late in the evening, slept at Mr. Gillespie's house. On the 26th we left Johnson's rancho for an examination of the banks of the Yuba river and the adjoining country, striking the Yuba at " Cordua's" or "Speeks's" bar, direction N. 25° W., distance seventeen miles. Here we found a company of nearly one hundred miners busily engaged in the bed of the stream, which they had partially exposed by constructing an oblique dam extending nearly across. They informed us that their operations at this point had proved extremely successful, each individual averaging from two to three ounces per diem. Deer creek joins the Yuba on the south about two miles above Cordua's bar, upon the banks of which the richest deposites of gold yet discovered in California have lately been found.

The result of our examination was the selection of a site about half a mile above the store at Johnson's rancho, on Bear creek. This point possessed all the merits of others examined, and was comparatively free from objectionable circumstances. Being at the foot of the Sierra Nevada slope, the reserve could be made to include a sufficient space of high ground, not subject to the periodical overflows of the creek, for building the necessary barracks for the troops, while the rich bottom land border

ing the creek on either side afforded an excellent opportunity for garden-
ing purposes. A grove of fine oaks answers the purpose of shade for an
encampment during the summer months, while the hills in the vicinity
are covered with trees, affording a sufficiency of wood for fires, and of logs
for houses. The summits of the adjacent hills crop out with a species
of sandstone, which, if quarried, would answer admirably for the construction of chimneys and foundations for buildings. Its proximity to the
rancho is also an advantage, insuring a constant supply of fresh meat,
and we found the water of the creek at all times healthy, cool and pleasant.
But the central position of this locality is probably its greatest advantage.
The Truckee emigrant route, over which was passing an average of one
hundred wagons and two hundred emigrants per diem, the wagon road to
the Yuba mines, that to the Feather river "dry diggings," the trail to Rose's
rancho on the Yuba, striking into Lawson's route at a distance of twenty
miles, and the paths to the Bear creek diggings, all intersect at this point;
while the post is, moreover, within a few hours' ride of all the principal
ranhos and Indian ranchoerias, or villages, in this part of the valley.
There are several points on Feather river, and one upon the Yuba near
Rose's rancho, where much prettier and more romantic sites may be found
combining most of the advantages of this upon Bear creek, but the notori-
ously unhealthy character of these locations offered an insuperable objection to their being selected.

We found, upon inquiry, that there had been but little sickness upon
Bear creek during the summer among the Indians or the emigrants, who
had encamped for weeks upon its banks; while at Sutter's farm, on
Feather river, and at the ranchos on Yuba, most of the occupants had
suffered with the periodical fever, and several deaths had ensued.

Within three or four miles of the post, gold is found in small quantities
in the ravines running towards the creek; digging for which will prove a
healthy and profitable recreation for the unemployed soldiers, and will probably be the occasion of preventing many desertions that might otherwise
occur.

We returned to our encampment on the 27th of September, and made
observations for the latitude and longitude. The longitude worked out
was probably incorrect, the chronometer having changed its rate from travelling in the wagon: the latitude we made 39° 0' 6". We found a very
unpleasant incident had occurred on the morning of the 27th. My servant
Manuel Montano, who had been with me some two months, and had
always appeared remarkably honest, faithful, and attentive, deserted, after
cutting a hole in the back of my tent, through which he thrust his hand
and removed from a box beneath the head of my bed my purse, containing $425 public funds, a gold pencil, gold watch and chain, a pistol,
and some other articles of value; he also cleared out his room-mates, taking
whatever money they possessed, and left on the best horse we had, carefully selecting the best saddle and riding equipments in the possession of
the party. We discovered this at reveille, when he had been gone probably
some hours, and Mr. Kemp immediately volunteered to start in his pursuit. I accordingly despatched him, with another horse, upon the road to
Sacramento City, judging that Manuel, who was a Chilian, and knew
nothing of the country, would probably take that direction with which
alone he was familiar. I have never seen either of them since. This unpleasant occurrence filled us all with surprise and consternation, and threw
a general gloom over our little party, which it required several days to

dispel. On the 28th we struck our tents at 8 a. m., and following the
course of Bear creek, passing in our progress numerous encampments of
emigrants, who had halted for the purpose of recruiting their cattle on the
fine high grass which abounded on the banks, we soon arrived at Johnson's rancho; passing which about a mile, we encamped on the creek
nearly at the centre of the present reserve. My party having been reduced
by Manuel's flight and Kemp's pursuit, I replaced them by hiring two
persons at the rancho—the one as cook and laborer, the other as an axe-
man. On the 29th I commenced the survey of the reserve, running the
base line as nearly parallel to the direction of the stream as possible, which
gave it the bearing N. 47° E., the perpendicular erected at each extremity having the bearing from the extremities of the base N. 43° W.,
each line one mile in length, and the base two hundred yards from and
parallel to the southeast bank of the creek. This gave the reserve of one
square mile, which was situated upon the northwest bank of the Bear
creek, with the exception of the strip two hundred yards in width upon
the other bank, reserved for the purpose of preventing individuals from
squatting or settling in the immediate vicinity of the post. Finding the
soil to be equally available for gardens and pasturage, Captain Day saw fit
to include a quarter mile upon the southeast bank. The lines were accordingly moved this distance to the southeast; preserving, however, their former bearing. I caused stakes to be driven at every furlong, with a board
strongly nailed upon each, and legibly marked "United States reserve."
    The variation of the compass I found to be 15° 30' east. We found
the heat most intense during the mid-day—so much so that I was obliged
to conduct the work merely during the morning and evening; but in
spite of this precaution my assistant and two employés were taken sick,
apparently from the effect of exposure to the sun, and, being confined to
their tents for several days, somewhat delayed the completion of the
survey. We made observations for the longitude and latitude nearly
every day, and found the latitude 39° 2'; longitude by the chronometer, repeated observations, 120° 45'; but I place but little faith in this,
for the unhappy instrument had been so jolted over the rough roads, and
had altered its rates in such a very unexpected and inconsistent manner
during the march, that we could place but little reliance upon its accuracy. On the 7th and 8th of October, I made an examination of Bear
creek for about 25 miles from the encampment. On the 9th the weather,
which had gradually for two or three days been growing damp and un-
comfortable, changed to rain, pouring down in torrents during the night
of the 9th, and continuing to rain heavily at intervals during the 10th.
Many miners came down from the upper diggings on the creek during
these two days, reporting that it was impossible to work on account of
the severity of the cold. This sudden change from the excessive heat
of the last week struck me as particularly extraordinary, even in this extraordinary country. On the 14th the weather, which had been pleasant
from the 10th, although much cooler in the mornings and evenings, again
changed, favoring us with a tremendous gale from the northwest, accompanied by squalls of rain, which created great havoc among the tents of
our encampment; but again all cleared away pleasant before the morning
of the 15th. These unexpected rains were somewhat alarming to us,
as, having the main part of our work to execute in the examination of
the Sacramento valley, we dreaded they might prove the immediate pre-

cursors of the rainy season, which, if commencing immediately, would have put a sudden termination to our labors. On the morning of the 15th, having finished the survey and made a map of the reservation, which I left with Captain Day for the use of the post, I broke up my encampment at Bear creek and marched north, for the purpose of examining that portion of the valley mentioned in my instructions.

I had previously dispensed with the services of my assistant, which I had found not entirely necessary, and quite inadequate to their expense to the government. We struck our encampment at daylight; but, owing to the many little matters of business necessary to be attended to before leaving, did not get fairly upon the road until 8 a. m. Keeping generally a north by west course, we crossed Dry creek, at a distance of three miles from our encampment. This in winter becomes quite a considerable stream, which, rising in the low hills, flows in a southwesterly direction and empties into Bear creek, about 5 miles below "Johnson's."

We arrived upon the southeast bank of the Feather river, six miles below the junction of the Yuba, at 5 p. m., at a point immediately opposite Sutter's "Hock farm"—here we encamped for the night. This is the most beautiful situation that I have seen in California. The river, which at this place is about six hundred yards in width, is lined on either bank with majestic sycamores, in a fine grove of which, upon the west bank, is situated Captain Sutter's farm-house, a remarkably neat adobe building, whitewashed and surrounded by high and well built walls enclosing out-houses, corrals, &c. There are about 100 acres of excellent land enclosed and cultivated upon the west bank, which yields the most astonishing crops of wheat with very little labor. The river is filled with salmon; and we observed two seines drawn across the river, about a mile apart, from which I was informed the occupants of the farm-house obtain a plentiful supply. About 200 yards above the farm-house is situated a rancheria of Indians, some 300 in number. This village consists of about 20 mud ovens, partly above and partly below the ground, with a small hole for egress in the side. They had just commenced the collection for their winter stock of acorns, and had many high baskets, containing probably forty or fifty bushels of this species of provender, standing about. They were mostly naked, and kept up a dismal howling all night as a tribute of respect to one of their number who had departed this life on the day previous. They were suffering much with the prevalent fever, as were the occupants of the farm-house, several of whom were sick at this time. On the 16th two of my party, Hyer and Flint, were seized with the fever. Having administered a prodigious dose of quinine to each, I proceeded to make an examination of the river in each direction from my encampment, which occupied the remainder of the party during the 16th and the 17th.

I determined the latitude of Sutter's farm 39° 2' 14", nearly identical with that of the post at Bear creek; longitude, by chronometer, 121° 14' 45". On the 18th, one of my patients having recovered, I left this beautiful but unhealthy river, having made a bed in the wagon for Hyer, who, although suffering much, refused to be left behind.

We crossed at the lower seine, about one mile below the farm-house, having much trouble in getting up the bank on the west side. The mules refused to pull until, by the assistance of some Indians, the baggage was entirely unpacked, when, by dint of much shouting, screaming, and profanity from the party, they managed to draw the empty wagon to the top. Passing the farm-house to the left, we now entered upon a fine, level prairie, the soil of which was of the richest description, and its surface dotted with the "long-acorn oak" for a distance of two or three miles from the river. We saw many antelopes and deer, but did not succeed in obtaining a shot at them. The Buttes were now in full view, presenting the singular spectacle of a range of mountains from one to two thousand feet in height, and some twelve miles in length, rising from the middle of a broad prairie, and entirely unconnected with any range. Keeping by compass the direction N. 45° W,, we travelled on until night-fall, when we arrived within three miles of the spurs of the north Butte. The heat had been intense, and having crossed the prairie without water, we were suffering extremely for the want of it, as were our animals, whose labor in pulling the wagon over the dry and cracked soil, which yielded at every step, had been very severe. I therefore caused the party to halt, and sent off three men in different directions to search for water, which they found and reported at the northern extremity of the Butte, to which place we proceeded at once and encamped, about 11 p. m., after a very fatiguing day's march.

The "Buttes," or, as they have generally been termed, the "Three Buttes," have been erroneously represented to be three isolated peaks rising in the prairie. They are in reality a range, containing some twenty peaks, and about twelve miles in length, by five or six in width. The two principal peaks are at the northern and southern extremities—the former 2,483 feet above the plain, the latter 1,841. The northern "Butte" is much the most remarkable. It is of a nearly conical shape, very steep, its summit cropping out with trap rock disposed in every variety of fanciful figures, one of which (the actual summit) is a tall turret shaped rock fifty-six feet high, and above twelve feet in diameter. The base runs off into five distinct spurs to the north, between each two of which lies a beautiful green valley, watered by streams having their source in never failing springs about midway up the mountain. The soil is of the richest alluvial deposite, giving no indications of minerals, and the whole range is studded with the holly oak, and thickly inhabited by almost every description of game—bear, antelope, black-tailed deer, cuyotes, and even the panther, or puma, (called the "California lion,") whose roar resounded about our encampment during the night with startling distinctness. We made the latitude of the north "Butte" 39° 12′ 32″, longitude 121° 28′ 36″; of the south, latitude 39° 9′ 32″, longitude 121° 29′ 6″. It was in one of the little valleys that we encamped, by the side of a beautiful spring of clear, cold water. Here we found many human bones, and the embers of a large fire, in which were the remains of a carpet bag or valise and some plates and cups. We observed, also, a newly made grave in the valley, with a cross placed at its head, on which had been an inscription, but it was now illegible. I succeeded in clambering to the summit of the north "Butte," from which I had an uninterrupted view of forty miles in every direction, observing distinctly the courses of the Sacramento and Feather rivers for at least that distance, and on the right and left the snowy mountains and coast range extending north and south as far as my eye could reach. On the 21st, two or three of my men declined proceeding on account of the day, it being Sunday. I had allowed them while engaged in the survey at Bear creek to rest on this day; but, as they were receiving very high wages for very small equivalents, I did not conceive it necessary to cease from travelling, particularly as I very much doubted their sincerity in the matter. I therefore informed them that I had not the slightest objection to halting every Sunday, but that, as they would do nothing on that day, I should infallibly stop their pay on every occasion of this kind. Upon hearing this their devotional feelings subsided with vast rapidity, and they professed themselves ready to proceed. I had been detained for two days at the "Buttes" by the sickness of two of my little party; but the bracing air of this delightful spot acted upon them like a charm, and by the 22d all were able to move forward. Taking a westerly direction, we marched across the plain, which was of rich alluvial soil, but cracked and parched by the heat, until we struck Butte creek at a point about twelve miles from its junction with the Sacramento, where we encamped. We saw upon the plain while crossing immense quantities of wild cattle, and a large drove of horses, containing over two hundred. These animals are, I believe, perfectly wild and unclaimed. We found it as difficult to approach them as the antelope and elk, which in some instances we observed feeding with them.

On the 23d and 24th we examined Butte creek, and the valley beyond, contained between it and the Sacramento. This beautiful stream has its rise in the western spurs of the Snowy mountains, and runs south with little variation in its course some 50 miles, where it makes a wide bend to the west in latitude 39° 9', and empties into the Sacramento. It is in many places of considerable width, but everywhere of great depth, carrying, I should imagine, as much water into the Sacramento as the "Yuba," the principal branch of the Feather river. Near its mouth it widens to about 600 feet, the ground in the vicinity being marshy and covered with tulé, and the banks difficult of access on account of the density of the alders and grape-vines with which they are lined. There are many clusters of beautiful trees—oaks, sycamores, and ash—upon its banks, but it is not thickly wooded, as is the case with the Sacramento and Feather rivers and their branches. The plain beyond is of rich alluvial soil, covered with fine grass, which was at this time almost dried up, upon which subsisted herds of wild cattle, horses, elk, and antelope. The "tulé" swamps do not extend far above "Butte creek;" there are but two or three isolated marshes of this description on the west bank of the Sacramento. There are two rancherias of Indians upon this bank, containing probably some 200, male and female. They subsist, like those already noticed, upon fish and wild grapes in the summer, and acorns and pulverized grasshoppers in the winter season, and appeared peaceably enough, but very disgusting to behold, being almost without exception stark-naked and excessively filthy.

Two deputations visited my camp on Butte creek: they behaved very civilly, and received and devoured some biscuit, which I administered to them, with a rapacity quite painful to behold. The men were all armed with bows and arrows, and the women furnished with baskets of a conical shape. I found out by signs that they were on a grape-gathering excursion. We found the creek to be plentifully supplied with a large white or grayish fish, weighing several pounds, and of a kind that I had never seen before. They resembled bass, and were delicious eating.

On the 25th we marched up the creek about 15 miles, and encamped near the crossing of Lawson's route. The stream here widens into a little bay, about 500 or 600 feet across. very deep, and about a mile in length. We made observations here, as we had done at our last encampment, at which we made latitude 39° 20' 26", longitude 121° 41' 15"; finding latitude 39° 31", longitude 121° 46' 15". The creek was at this place almost literally covered with ducks and brant, of which we shot many. We made an observation at sunrise for the variation of the compass, which was found to be 16° 49' east; a very large variation, but not as great as at the Buttes, where we found it nearly 18°. I was disposed to attribute this to local causes or attraction, but could find no evidence of iron, or in fact any other mineral, in the soil. We observed this day, as before, numerous herds of cattle and horses, one of which came to water at the creek about sundown, within a few hundred yards of our encampment. We observed comparatively few troublesome insects in this part of the country. Fleas, so annoying in the lower settlements, are here almost entirely unknown. Their troublesome office is, however, disagreably supplied by an unpleasantly-smelling pismire which covers the soil, and by its nocturnal rambles effectually banishes sleep. We left Butte creek on the 26th, and after travelling south of east about 15 miles over the dry and parched prairie, which was cracked in some places to the breadth of six or eight inches, rendering the wheeling extremely laborious, we struck "Lawson's route." This we found an extremely good road, upon which we marched some 12 miles further, passing many emigrant-wagons filled with dirty and unhappy-looking women and unwholesome children, and encamped on the bank of Feather river, six miles above the ford. The eastern valley, between Butte creek and the spurs of the Sierra, is from 30 to 40 miles in width, of rich soil, and covered during the spring and summer with fine grass. It is watered by the Feather river and its branches. This river rises in the Sierra, and flows slightly west of south through the valley, emptying into the Sacramento. It is remarkably straight throughout its whole course, making four small or abrupt bends, and is generally in the same latitude; wider, though not as deep as the Sacramento. Its banks are thickly wooded, for some two miles in depth, throughout its entire extent, with the holly and long-acorn oaks, sycamores, beech, ash, and alder trees. Its general depth during the dry season, or from the last of May until the first of November, is from two to ten feet; and its bed is much obstructed by sand-bars, which, while rendering fording at numerous points perfectly safe and easy, prevents entirely its navigation even by the smallest class of vessels. It is fed by numerous small creeks which run down the ravines of the Sierra, and in whose dry beds during the summer rich deposites of gold have been and still continue to be found. I observed three rancherias of Indians upon its banks within 12 miles of the crossing of "Lawson's route," which may contain in all from 300 to 500. They are all of the same wretched class with those observed upon the Sacramento. They appear perfectly harmless and remarkably good-humored, and many of them are in the employment of the emigrants who have squatted in the vicinity. On the 27th we marched six miles down the Feather river, crossing in about two feet water. A small island in the middle of the stream renders the ford more easy, and would be useful in the construction of a bridge at this point, which the high banks render practicable. We kept the river road for several miles, and then crossed the plain to the "Yuba," nine miles further. This plain is precisely of the same character as the others—unbroken, like them, save by a few ridges of low hills, and of the same rich alluvial soil. We encamped on the "Yuba," in the vicinity of Rose's rancho, a beautiful site for farming or grazing, but apparently subjected to but little cultivation. We found here a small adobe house, redolent with the odor of whiskey, and festooned with strings of jerked beef. The "Yuba" is a small but rapid stream, flowing southwest from the mountains, and emptying into Feather river. Its bed is rocky, giving its waters a turbulent character, particularly when swollen by the rains or melted snows of the Sierra. Its banks are wooded and not as high as those of Feather river, and it is occasionally subject to overflow. The soil in the immediate vicinity is of the richest description, and in the upper part of the river and its tributary, "Deer creek," the richest deposites of gold have been discovered. Leaving "Rose's rancho," we marched across the plain to Bear creek, striking the road from Noyes's rancho to Johnson's at a distance of about six miles from Rose's. We found Captain Day's command comfortably established at camp "Far West," upon the reserve, and preparations being made for building. We remained here during the 29th and 30th, and made observations for the rate of the chronometer, which, as I had supposed, had changed its rate very much from being jolted in the wagon.

The weather had prevented our making but one lunar observation, from which I established the longitude of the Buttes, and my map is projected from that data. Bear creek is about forty miles in length; it has two branches, which unite about thirty miles from its mouth; its course is nearly straight, as is the case with nearly all the rivers of the valley; and it empties into Feather river, being the second branch of importance in point of size. Its banks are thickly wooded towards its mouth, mostly with shrub-oak, buck-eye, and alder. In the summer it has but little water, but is never entirely dry; in the winter it becomes a deep and rapid stream, overflowing its banks to a very considerable extent. I have been informed that one instance has occurred of individuals going from "Johnson's rancho" to Sutter's fort in a whale-boat, the entire plain, forty-five miles in extent, having been submerged during a freshet; but I do not vouch for the truth of the statement. Leaving camp Far West on the 31st, we travelled down the creek to its mouth, then continued upon the branch of the Feather river until we had arrived within a mile of the town of Vernon (situated at its mouth, and supposed to be the head of navigation upon the Sacramento,) where we encamped. Feather river, near its mouth, is a very broad and beautiful stream. Its banks are heavily timbered, and some fifty feet in height, coming down abruptly to the water. There is a sufficient depth of water as far as the mouth of Bear creek to float any small-size vessel; but the frequent occurrence of extensive sand-bars renders the navigation to this point at present impracticable.

During the night we killed three veritable raccoons, the first that I had ever seen in this country. The town of Vernon is situated at the junction of the Feather and Sacramento rivers, and that of Frémont immediately opposite. Each contains some twenty houses and one or two hundred inhabitants. The valley of the Sacramento, on the western bank of that stream, is for the most part a barren plain, with little vegetation or water. It is from thirty to forty-five miles in extent, being bounded on

the west by the coast range of mountains.   There are no streams emptying into the Sacramento from the west, south of latitude 39½° north, with
the exception of "Puta" and Caché creeks.   The latter is the outlet of
Clear Lake, flowing from its southeast extremity, and losing itself in the
"Tulé" swamp which borders the western bank of the Sacramento about
six miles southwest of the mouth of the Feather river.   "Puta" creek
rises in the coast range, and, flowing southeast, empties into the Sacramento about fifteen miles below the mouth of the American river.   The
whole country between the creeks is liable to overflow, and is very dangerous to attempt travelling after a heavy rain.   The "Tulé" swamp,
upon the western bank of the Sacramento, extending to the vicinity of
"Butte" creek, and occurring occasionally above, is from three to six
miles in width, and is impassable for six months out of the year.   A cor-
duroy road may, however, be made over it, which has been used in some
instances with success at all seasons.

There is a short road of this kind in the rear of Sutter's, which I believe has proved entirely successful.

We reached Sacramento City on the 2d of November, and encamped
upon the outskirts during a tremendous storm of wind and rain, which
proved to be the *bona fide* commencement of the rainy season.   Having
paid and discharged my party with the exception of a teamster and Mr.
John Day, whom I retained as assistant, and finding it impossible to sell
my animals and wagon to advantage, in consequence of the identical
reason that prevented my purchasing, (*i. e.* the expedition for the relief
of the suffering emigrants which had just preceded my return, and whose
mules and other property were being disposed of at auction,) I concluded
to take them all to "Pueblo de San Jose," where I had reason to believe
they would meet with a ready sale.   I accordingly crossed the Sacramento on the 6th, at 2 p. m., and encamped on the verge of the Tulé,
being unwilling to run the risk of remaining all night upon the plain beyond.   It rained tremendously all night, and we were glad to take the
road at daylight in the morning.   We crossed the Tulé safely, but found
the road beyond extremely heavy and covered in some places with water.
These we passed without difficulty, observing to keep carefully in the
road, which being packed by travel, was not as dangerous as the plain
upon each side.   We had arrived within half a mile of "Puta" creek,
where I observed with astonishment and alarm that a strong current was
setting down the road, and that the water was deepening around us with
rapidity.   I at once comprehended that the creek had overflowed its
banks, and that we were in a dangerous position.   I gave the order
to the teamster to turn about immediately, but it was too late—the
mules sank at once on turning from the road, the wagon was fast
blocked in the yielding mud, and the water, as we afterwards found, was
gaining on us at the rate of four feet an hour.   It was with the utmost
exertion and no little danger that we succeeded in getting the mules extricated from the wagon, from which I had already saved my chronometer and best sextant, my drawing instruments and papers.   We were
compelled to abandon the wagon, with the remainder of the instruments
and all our personal property, and return to Sacramento City, where I disposed of my animals at auction.   The whole of the valley, I found upon
my return, had been made a perfect quagmire by the recent rains, and
several wagons had been lost on the same day in the attempt to go to
Frémont from Sacramento City.   I arrived at Benicia on the 10th of

November, from which place I had the honor to send you a detailed report of tne loss of my wagon, and other circumstances connected with my expedition. From the above general summary of the journal kept while engaged upon the expedition, and from the accompanying map, an idea may be obtained of the geographical position of that portion of the valley mentioned in my instructions, and, as far as the means of communication, its capability of supplying forage, and its agricultural character are concerned, of its *general* and *military* resources. It will be seen that there is no point in this portion of the valley at which a military post can by any possibility be required, that is absolutely free from objection. That already selected combines as many advantages in a military point of view as it is possible to obtain. A position at the north Butte, or upon Butte creek, at some point near the crossing of Lawson's route, would be undoubtedly more pleasant in every respect, perfectly healthy, and would possess every requisite for the comfortable subsistence and shelter of the troops; but beyond the attaining of these points, I can conceive of no advantages to be derived from establishing a post in that vicinity. There would be no inhabitants to protect, and nothing in fact to protect them from. As far as regards *building materials*, all points are equally eligible. The soil of any portion of the valley mixed with water and chopped grass, and exposed to the sun, makes excellent adobe or sun-dried brick, which is probably more economical, easier of working, and better adapted to the climate, than either timber or stone. Pine timber may, however, be obtained of the best quality upon the table-lands at a distance of about twenty miles from the commencement of the hills; the oak, which forms a greater portion of the timber upon the streams, being too hard and brittle to work easily, and soon decaying when exposed to the weather. Two steam saw-mills are now in process of erection upon Bear creek, about four miles above the newly established post, from which, when in operation, sufficient supplies of pine timber may be obtained.

There are two rancherias of Indians upon the Sacramento, one upon Butte creek, three upon Feather river, and one upon Bear creek (about ten miles above the post,) which have come under my observation. All of these together must contain something under one thousand individuals, men, women, and children. I was informed that upon the Upper Feather and Yuba rivers were some two or three thousand living in the hills, but whether belonging to the same tribe with those of the valley I could not ascertain. All that I have seen appear equally ugly, harmless, and inoffensive; but, being perfectly barbarous, and acting, as I imagine, more from instinct than reason, they are liable to commit, at any moment, some unexpected outrage, for which neither themselves nor any one else can assign a reason.. I was informed that the rancheria upon the Sacramento had, within a few months, committed three murders upon white men travelling upon the western bank; I was not able, however, to arrive at any details, and am not satisfied that the report was worthy of credence.

As far as I could ascertain, by inquiry, from those persons most likely to be best acquainted with their character and habits, all of these Indians are to be viewed with suspicion and distrust; and I found it generally conceded that those termed "Christian Indians," who, by their intercourse with the whites, had added to their original qualification of low cunning some gleams of intelligence, were by far the most dangerous, being invariably found to be the ringleaders in all thefts or other outrages committed by a rancheria. Should the present rapid emigration to this country continue during the ensuing year, the entire valley will undoubtedly be thickly settled with a hardy population, who, attracted by the fertile soil and beautiful scenery of the banks of Feather river and its branches, will brave its sickly climate, preferring to reap a sure and lucrative harvest from agricultural pursuits to enduring the hardships, exposure, and sickness of the mines for a doubtful prospect of immediate wealth. In this case the post at Bear creek, instead of being, as now, on the frontier of civilization, would be surrounded by a population perfectly able and willing to help themselves; and it might become advisable to establish a station either further to the north, in the valley, or in advance, upon the "Truckee" road, according as either became finally the main route for emigration.

I have the honor to enclose with this memoir a map and copy of the reserve at Bear creek; also, a map and copy of the Sacramento valley, from the American river to the mouth of the Butte creek, which embraces that portion of the country referred to in my instructions.

Trusting that they may meet the approval of the commanding general, I have the honor to remain, sir, with high respect, your obedient servant,

GEO. H. DERBY,
Lieutenant Topographical Engineers.

Major E. R. S. CANBY,
Adj't Gen., Tenth Military Dep't.

———

HEADQUARTERS THIRD DIVISION,
Benicia, February 27, 1850.

COLONEL: I have the honor to transmit the report of Brevet 2d Lieutenant R. S. Williamson, of the topographical engineers, of the reconnoissance made by the late Brevet Captain W. H. Warner of a route through the Sierra Nevada by the Upper Sacramento. It will be seen that, from the waters of the bay of San Francisco, following the valley of the Sacramento on the west side, to the foot of the hills at the base of the Sierra, is a smooth inclined plane, without obstacle. Following up Pit river, or Cow creek, and then crossing to Pit river, a route can be carried through to the eastern foot of the Sierra, with a gradient of thirty-eight feet per mile—thus passing from the Pacific to the eastern base of the Sierra Nevada almost without any of the obstacles common to most of the roads in the United States.

Captain Warner's death cannot be too deeply lamented. He was devoted to his profession, and well qualified, by his perseverance, habits of endurance, and his uncommon accuracy, to reach an eminent position in it. He and Lieutenant Williamson were subjected to great hardship and suffering from sickness and the want of wholesome food, and their labors were performed under great disadvantages. Lieutenant Williamson deserves great credit for the manner in which he has concluded the work and put on record the valuable result of their reconnoissance.

With respect, your obedient servant,

PERSIFOR F. SMITH,
Bt. Major Gen commanding Division.

Bt. Lieut. Col. WM. G. FREEMAN,
Ass't Adj. Gen., headquarters of the army.

SONOMA, CALIFORNIA, *February* 14, 1850.

SIR: Orders from the headquarters of the Pacific division, dated June 27, 1849, were issued to Captain W. H. Warner, of the topographical engineers, to take charge of an exploration from the upper Sacramento, across the Sierra Nevada, to Humboldt river; the main object of the expedition "being to discover a railroad route through that section of the country." At the very moment it had been fully demonstrated by Captain Warner that such a route across the Sierra Nevada was practicable, and when nearly all the required information had been obtained by him, he was murdered by the Indians. In consequence of this lamentable occurrence, the command of the topographical party devolved upon me, and it now becomes my duty to submit to you, for the information of the major general commanding the division, the following report, which imbodies as much of the information collected by Captain Norris as came to my knowledge.

Four commissioned officers and eighty men were ordered to compose the escort of Captain Warner, and the party was to move from Benicia on the first of July, with supplies for four months. The detail for the escort was not received by the officers ordered till some days after the first of July, and the collection of the different detachments composing it, with their preparation for the transportation of the supplies, &c., consumed so much time, that the escort did not start from Benicia till about the first of August, arriving at Sacramento City on the third. The escort was commanded by brevet Colonel Casey, of the 2d infantry, who was accompanied by Lieutenants Schureman and Gardner, of the same regiment, and Dr. Hewitt, of the medical staff. Captain Warner and myself, with three men, left Benicia on the 28th of July, arriving in Sacramento City and Sutterville on the 31st. Here Captain Warner hired a party of ten men, and also engaged the services of François Bercier, (ordinarily called Battitu,) as guide. This man was a very intelligent half-breed, born on the Red river of the north; had served some time with the Hudson's Bay Company in Oregon; had trapped in the rivers of the country we were about to explore, and proved himself to be a very valuable man for the short time his services were given to us. On the 13th of August we commenced our march up the Sacramento valley, on the east side of the river. Our party consisted of eleven hired citizens and two soldiers, and we carried with us a light wagon, a cart, and a few pack-mules. As there was no need of the escort until we should be among the hostile Indians of the mountains, we did not delay our march in order to keep company with it, but advanced to Lassen's (universally pronounced Lawson's) rancho, on Deer creek, where we remained a few days to jerk beef. Captain Warner here gained a great deal of information with regard to the route across the mountains, from Mr. Lassen, who had guided a party across the country from Missouri in the summer and fall of 1848; had deviated, near the great bend of Humboldt river, from the previously travelled road, and after much difficulty succeeded, with great good fortune, in finding a continuous ridge from the summit of the Sierra to the Sacramento valley just at the place where his own rancho was situated. He represented this to be an excellent road after the first ten miles, and expressed his opinion that it was a practicable railroad route. This induced Captain Warner to strike into the mountains from this point, instead of proceeding further up the Sacramento valley. The escort

Part ii—Ex.—2

arrived at Lassen's three days after our party, and it was found that it was impossible to proceed into the mountains with the whole command, and move with sufficient rapidity to return before the snows commenced. Captain Warner, therefore, made a requisition upon Colonel Casey for a detachment of a commissioned officer and twelve men, which, with the citizens he had hired, would make his party sufficiently strong, as he thought, to resist any attack that might be made upon him by the Indians. Animals were required from Colonel Casey to pack our provisions and mount the infantry detachment, and this left him with very small means of transportation. Captain Warner had, by this time, given up all idea of being able to reach Humboldt river, and he proposed to follow the Lassen trail till he struck Pit river, then follow up this river to its source, and endeavor to find a pass through the eastern ridge of the Sierra near that point. Colonel Casey, in the mean time, was to advance slowly with the provisions, and was expected to be on Pit river to meet us on our return, and furnish us with supplies to enable us to explore a route either down that river or Cow creek. On the 24th of August we left Deer creek and entered the mountains, our party being about thirty-four strong, and having with us the wagon, cart, and about eighty-six animals. We found the hills quite steep, and the road very rough and rocky, so that it was with much difficulty we made nine miles, our mules being very poor, though the best that could be obtained. The next day, though the loads of the wagon and cart had been much reduced, we marched but nine miles farther, the wagon not arriving at camp till after dark, and the mules in the cart giving out entirely. We were exceedingly desirous of carrying the wheeled vehicles with us, as we wished to continue to measure the road with the odometer, but they retarded our march so much that Captain Warner determined to send them back, although it was practicable to take them along at a very slow rate. Our men were here taken sick hourly, and the next morning nearly half the party was unfit for duty. Captain Warner had had a fever for several days, and was now quite sick, and in fact it was impossible for us to advance immediately. During our second day in the mountains we met several men from the United States, and learned, to our surprise, that they were the advance of from ten to twenty thousand emigrants, who were coming into the country on this trail. The advance parties were composed mostly of 'men who had left their trains, under a false impression as to the distance to the valley, and they were in a starving condition. They came to us in such numbers, begging for provisions, that our supply was materially diminished. As soon, however, as we met the trains, we found the emigrants generally pretty well provided. Captain Warner succeeded in hiring several of them to supply the places of those of our party too sick to advance; and leaving seven men in charge of the surgeons, and sending back our wagon and cart by one of the emigrants, with everything that we could dispense with, we left our camp on the fourth of September, and on the evening of the fifth we reached the headwaters of Deer creek, about fifty miles from Lassen's. Here we were obliged to wait another day, and to leave behind five more men, too sick to proceed. As the detachment of soldiers under Lieutenant Gardner was now reduced to ten men, and as he himself was unwell, he remained in camp with the sick. It is a singular fact that, during the first ten days in the mountains, not a single one of our original party escaped an attack of fever. Upon leaving, the headwaters of Deer creek, we continued to follow the trail, crossing the headwaters of Feather river, striking Pit river on the 13th of September, and reaching the headwaters. on the 18th, which are near Goose lake, a body of water over twenty miles long, and impregnated with salt. The impression produced by Mr. Lassen, as to the character of the road, was decidedly incorrect. For thirty-six miles after entering the mountains, the road was very rough, stony, and steep, continually ascending till we had attained an elevation of over five thousand feet. The road followed a ridge between Mill creek and Dry creek, which streams flowed in steep ravines on either side, into which it was necessary to descend to get water. We seemed at the end of this thirty-six miles to have arrived at the summit of a continuous range of mountains, from which we descended about two thousand feet in the next ten miles. It appeared certain that no railway could be constructed over the road we passed, unless extensive inclined planes were used. Between this mountain range and the range in which Pit river rises, the country is mountainous, but interspersed with numerous valleys from five to twenty-five miles in length, and between three thousand and four thousand feet above the Sacramento valley at Lassen's. The mountains near the source of Pit river are between six and seven thousand feet above Lassen's, and, as far as we could ascertain, formed a continuous unbroken range. This range had been considered by Captain Warner as presenting the greatest obstacle in his way; and as it was impossible for any railroad to cross it where the Lassen trail crossed, he determined to pursue his examination to the northward, in hope of finding a depression in the mountains. As many of our mules were worn out, he determined to take a small picked party, leaving the rest with me in camp; and accordingly he started on the 20th September with nine men, including the guide, and ten days' provisions. He passed to the northward, leaving Goose lake and the mountains on his right, and, after travelling seventy miles, passed through a deep depression in the mountains, which differed very little in altitude from the valley of Goose lake. The result from my calculation of the barometric observations gives the altitude of the highest point passed over by Captain Warner but one hundred and seventeen feet higher than over camp near Goose lake. It is true that a difference in the state of the atmosphere, at the times of observation, may have caused an erroneous result; but as no such difference was apparent, the error could not be great. The discovery of this pass, then, makes a railroad route perfectly practicable, as far as the eastern or main spur of the Sierra Nevada is concerned. After going through the pass, there is no impediment until the road reaches the point where our trail first struck Pit river. This river, which is the main fork of the Sacramento, is said to have taken its name from the numerous pits on its banks, which are sometimes twenty feet deep, and which had been dug by the Indians as traps for deer, bear, &c. It is, as far as we followed it, a slow, sluggish stream, descending about eight feet in the mile, and having ample room between its banks and the hills for the construction of a good road. The banks in a few places showed signs of being overflowed in the spring, but generally appeared high enough to confine the waters. The hills near the banks were not cut up by ravines, and we observed no dry beds of streams near the banks which would require to be bridged. The hills sometimes approach very near the bank and sometimes diverge, forming fine valleys. Where the trail to Lassen's leaves the river, it strikes into steep hills, where it would be very difficult to construct the railway. In the opinion of Captain Warner, the road should follow Pit river, or, by crossing over to the headwaters of Cow creek, follow down that creek to the Sacramento valley. The altitude of the point where the road leaves the river is three thousand eight hundred feet above Lassen's; and as the distance to the valley must be at least a hundred miles by either of these routes, there would be an average descent of thirty-eight feet to the mile, which is perfectly practicable for a railroad route. Which of the two routes above mentioned is preferable, can only be determined by a minute examination. Our guide, Battitu, said he had travelled down Pit river to the valley, and after passing the cañon of twenty miles, the banks were open. From what I saw of the cañon, I think a road can be cut through it with much less labor than is expended upon many of our eastern roads. The whole country through which the road would pass till it reaches the valley is well timbered with pine, fir, and cedar; and abundance of burr or mill-stone is found, as well as another silicious rock, well adapted for road covering, which would make a good building-material for common purposes. Some specimens of these stones, taken from different locations on Pit river, I present for your examination. When Captain Warner had discovered the pass, and reached the eastern base of the range, he travelled to the southward, intending to recross the mountain on the Lassen trail. On the 26th day of September he was riding in company with the guide, a short distance ahead of his little party. They had descended a little ravine and were ascending the rugged hill on the other side, when a party of about twenty-five Indians, who had been lying in ambush behind some large rocks near the summit, suddenly sprang up and shot a volley of arrows into the party. The greater number of the arrows took effect upon the Captain and guide, and both were mortally wounded. The Captain's mule turned with him, and plunged down the hill; and having been carried about two hundred yards, he fell from the animal dead. The guide dismounted and prepared to fire, but finding he could not aim his rifle, he succeeded in mounting and retiring down the hill. He died the next morning. The party were thrown into confusion and retreated at once. Two men, George Cave and Henry A. Barling, were badly wounded. Cave died before reaching the valley, while Barling reached Benicia, was placed in the United States hospital under charge of Assistant Surgeon Deyerle, and has now nearly recovered. Captain Warner's body was visited several times, and his note-book, &c., brought to me. The Indians who made this attack are supposed to be of the same tribe, and have the same manners and customs, as those in the immediate vicinity of Tlamath lake. They caused a great deal of trouble among the emigrants by stealing their cattle in the night; and they acted with a great deal of caution, never showing themselves during the day. They have no other arms than bows and arrows, and generally go entirely naked. They seemed to have been emboldened by the presence of so small a party so far from the emigrants' trail, and presented themselves in considerable numbers in the vicinity of Captain Warner's camp for several days preceding the attack. It is difficult to make an estimate of their numbers, but they certainly can form a formidable body. I met on Pit river a party under charge of Mr. Peoples, who had been sent by Major Rucker, with a supply of provisions to the relief of the emigrants. He told me that several of his men, while hunting among the hills for some lost animals, had discovered a body of the Indians, which they estimated at five hundred strong.

The information contained in Captain Warner's note-book was so full that it would have been superfluous for me to have visited the pass he discovered. The lateness of the season, and the grass on the road having been nearly consumed by the animals of the emigrants, rendered it impossible for me to advance to Humboldt river. My party was not sufficiently strong for me to attempt to punish the Indians; and our provisions were nearly exhausted. For these reasons I deemed it necessary to return immediately to Lassen's. I sought for conversation with the most intelligent of the emigrants to ascertain the character of the road between the mountains and Humboldt river, and they were unanimous in asserting that there existed no serious impediment to the construction of a railroad at any part of it. Upon the return of the remainder of Captain Warner's party with their sad intelligence, I succeeded in making arrangements with some emigrants for the transportation of the two wounded men, in wagons, to Lassen's; and, this being done, prepared to return with the remainder of the party. I fortunately succeeded, a few days afterwards, in purchasing some provisions from an emigrant, which kept us scantily supplied till we reached the valley, which we did on the 14th of October. On the 10th and 11th we had the first snow of the season. I was exceedingly anxious to return by Cow creek; but, having no guide, and being so short of provisions, I thought it necessary to return by the same route we came. Colonel Casey, who, with the escort, had reached the headwaters of Deer creek, and had remained there for some time in camp, had preceded me a few days to Lassen's, and left that place on his return to Benicia. I here sent back three men to carry provisions to the wounded, and to bring them to Benicia. I left, for their transportation, the wagon which Captain Warner had sent back; and, with the cart and pack train, left Lassen's on the 16th, crossed the Sacramento near the mouth of Deer creek, and returned to Benicia on the western bank of the river.

I have the honor to present, with this report, a copy of the meteorological and astronomical observations taken upon the expedition, and a sketch of our route. I will state the data from which this sketch was made, to show the degree of accuracy of its different parts. In going from Benicia to Sacramento City, from there to Lassen's, and thence back to Benicia, the distances were measured with an odometer, and the courses taken with a prismatic compass. At each turn of the road a new course was taken and the reading of the odometer noted, while at the same time a sketch of the country passed over was made, and the position of houses, rivers, &c., determined. Hence I had complete compass notes, with distances actually measured, which I plotted. The positions of Sacramento City, Benicia, and Lassen's were then assumed on the map, by their latitudes and longitudes, and neither of the distances between the points so determined differed two miles from the corresponding distances as plotted; a degree of accuracy which is the result of great care and attention on the part of Captain Warner. From what I have above stated, I think the position of the different points on our route, while in the Sacramento valley, are laid down with accuracy. I have studiously avoided putting down anything that did not come immediately under our observation—confining myself entirely to a sketch of the route, except in the single case where I have taken from a map made by Captain Warner that part of the Sacramento river below Sacramento City, together with the bays of Suisun, San Pablo, and San Francisco. The numerous streams running from the Sierra Nevada to the Sacramento were all dry at the season we crossed them, excepting those with names; but in the winter season are formidable rivers, which would require to be bridged if the road was made on that side of the river.

There are, however, but two streams above Sacramento City, which come from the coast range. The part of the route from Lassen's to Pit river is laid down only approximately, as the road was so winding, and the forest so thick, that it was impossible to get our course with accuracy. After reaching Pit river, however, we had the nearly open bank to travel upon; and our courses were again observed with care, and the distance estimated by noticing the time taken to pass over any course. The latitude of our camp 29, near Goose lake, was ascertained from Captain Warner's observations, while its longitude was assumed approximately, by reference to Fremont's map.

As the Lassen trail is now an emigrant trail from the United States, a list of distances from camp to camp, as far as we proceeded on it, may be useful, and I therefore submit it with the map.

I have the honor to be, very respectfully, your most obedient servant,
R. S. WILLIAMSON,
Lieutenant United States Topographical Engineers.
Bvt. Lieut. Colonel J. HOOKER, •
Assistant Adjutant General, Pacific Division.

Meteorological observations taken during Captain Warner's explorations in the Sierra Nevada, 1849.

Date.	No. of camp.	Location of camp.	Barometer. Upper reading. Cent.	Barometer. Lower reading. Cent.	Thermometer. Attached.	Thermometer. Detached.	Time of observation.	Wind. Direction.	Wind. Intensity.	Altitude above Lassen's, in feet.	Remarks.
1849. Aug. 21	12	Deer creek.	39.61	35.78	30.7	78	Sunset	S. & W.	3		Smoky.
22	12	Near Lassen's.	39.63	35.98	16.2	60	6h. 30m. a. m.		0		do.
22	12	do.	39.61	35.82	25.7	73	Sunset	S. & W.	2		do.
22	12	do.	39.69	36	14.7	53	Sunrise				Cloudy and very smoky.
23	12	do.	39.63	35.91	21	71	Sunset		0		do.
23	12	do.	39.76	36.01	17.2	66	6h. 30m. a. m.	S. & W.	2		Clear; slightly smoky.
23	12	do.	39.83	35.94	34	96	1h. p. m.	S.	2		do.
24	12	do.	39.72	36.09	35	78	Sunset		0		do.
24	12	do.	39.77	36.04	16.5	63	6h. a. m.	S.	2		Clear.
25	12	do.	39.91	36.06	33.9	94.5	Noon	N. & W.	1		do.
25	12	do.	39.78	36.08	27.9	79	Sunset	N. & W.	1		do.
26	12	do.	39.82	36.07	19.4	68	6h. 30m. a. m.	N. & W.	1		do.
26	12	do.	39.96	35.91	34	95	Noon	N. & W.	0		Clear; slightly smoky.
26	12	do.	39.81	36.06	26.5	79	Sunset	N. & W.	1		Clear.
27	12	do.	39.79	36.86	18	64.5	6h. 30m. a. m.		0		Clear; slightly smoky.
27	12	do.	39.69	35.88	97.2	61.5	Sunset	N. & W.	1		Smoky.
28	12	do.	39.58	35.81	15.3	61	7h. a. m.	N. & W.	3		do.
28	12	do.	39.69	35.72	32	92	Noon	N. & W.	3		Clear.
28	12	do.	39.58	35.88	21.2	90	7h. a. m.		0		do.
29	12	do.	39.61	35.90	15.5	70	6h. 30m. a. m.				do.
30	13	Deer creek.	39.60	34.83	30.2	59	Sunset		0		Smoky.
30	13	Mill creek.	38.64	34.95	19.6	66	6h. 30m. a. m.		3		do.
31	14	do.	38.65	33.52	96.4	95	Sunrise	S. & E.	0		do.
Sept. 1	15	Dry creek.	37.44	33.42	26.8	69	Sunset		0		do.
1	15	do.	37.25	33.42	30.5	79	Noon	S. & W.	2		do.
2	15	do.	37.23	33.23	29.9	88.5	Sunset		0		Very smoky.
2	15	do.	37.12			85					

Meteorological observations—Continued.

Date.	No. of camp.	Location of camp.	Barometer. Upper reading. Cent.	Barometer. Lower reading. Cent.	Thermometer. Attached.	Thermometer. Detached.	Time of observation.	Wind. Direction.	Wind. Intensity.	Altitude above Lassen's, in feet.	Remarks.
1849. Sept. 3	13	Dry creek,	37.16	33.41	35.2	78	Sunrise	S. & E.	4	....	Very smoky.
Sept. 4	15	....do	37.14	33.46	21.7	74	...do	....	0	....	...do.
Sept. 5	16	One mile beyond summit of western range of Sierra	33.73	29.42	8.2	45	....do....	....	0	....	Clear.
Oct. 12	47 same as 16.	....do	33.66	29.66	4	....	6h. 30m. a. m.	Variable.	1	....	...do.
Sept. 5	17	Deer creek,	34.51	30.72	19.8	66	Sunset	....	0	....	Smoky.
Sept. 6	17	....do	34.48	30.80	2.4	36	6h. 30m. a. m.	....	0	....	...do.
Sept. 6	17	....do	34.57	30.77	30	85.5	Noon	N. & W.	1	....	Clear.
Sept. 6	17	....do	34.50	30.76	24	72	Sunset	....	0	....	...do.
Sept. 7	17	....do	34.28	30.78	0.6	39	Sunset	....	0	....	Smoky.
Oct. 11	46 same as 17.	....do	34.31	30.56	3.4	....	6h. p. m.	S. & W.	1	....	Clear.
Sept. 12	17	....do	34.46	30.72	4.4	....	6h. 45m. a. m.	N. & E.	1	....	Smoky.
Sept. 7	18	South fork of Feather river.	34.14	31.08	21.9	70	Sunset	N. & W.	1	....	...do.
Sept. 8	18	....do	34.12	30.66	22.5	36.5	Sunrise	....	0	....	Very smoky.
Sept. 9	19	North fork of Feather river.	33.87	30.12	21.6	70.5	Sunset	....	0	....	Clear.
Sept. 9	19	....do	33.72	30.90	1.6	34	Sunrise	....	0	....	...do.
Oct. 9	44 same as 19.	North fork of Feather river.	33.79	29.69	8.2	....	Sunset	Variable.	2	....	Foggy.
Sept. 10	19	....do	33.76	29.65	0.2	....	Sunrise	S. & W.	2	....	...do.
Sept. 9	20	Small lakes	33.03	30.61	18.3	62	Sunset	W.	2	....	Smoky.
Oct. 8	43 same as 20.	....do	33.89	29.42	2.4	23	Sunrise	....	0	....	Very smoky.
Sept. 9	20	....do	32.96	29.12	10.7	....	Sunset	S. & E.	2	....	Cloudy.
Sept. 10	21	Round valley	32.76	29.03	0.8	....	Sunrise	N. & E.	1	....	Foggy.
Sept. 10	21	....do	33.23	29.36	14.6	49	Sunset	....	0	....	Smoky.
Sept. 11	21	....do	32.94	29.47	8.8	13	Sunrise	....	0	....	Very smoky.

Date	Locality				Time	Wind			Remarks
Oct. 7	42 same as 21	33.34	29.14	13.5	Sunset	S. & W.	2	......	Cloudy.
	21 ...do...	33.23	29.14	0.5	7h. 15m. a.m.	S. & W.	1	......	do.
Sept. 11	23 Spring in mountain	33.04	29.54	17.4	5h. 45m. p.m.		0	......	Smoky.
Sept. 12	23 ...do...	32.99	29.34	3	Sunrise	S. & W.	1	......	Cloudy.
Oct. 6	41 same as 22	33.47	28.99	13.2	Sunset	S. & W.	1	......	Slight cloudy.
Oct. 7	22 ...do...	33.34	29.06	0.5	7 a.m.	do.	2	......	Dark in east.
Sept. 12	23 Branch of Pit river	34.59	30.92	23.5	5h. 25m. p.m.	W.	1	......	Cloudy and smoky.
Sept. 13	23 ...do...	34.50	30.88	3	6 a.m.		0	......	
Oct. 5	42 same as 23	34.88	30.85	10.8	Sunset	N. & E.	2	......	Very smoky.
Oct. 6	23 ...do...	34.75	30.87	4 6	Sunrise	N. & W.	4	......	Cloudy.
Sept. 13	24 Pit river	34.55	31.10	14	Sunset	W.	2	......	Very smoky.
Sept. 14	24 ...do...	34.69	31.17	1.3	6h. 30m. a.m.		1	......	Thick fog.
Oct. 4	39 same as 24	34.88	30.98	9.9	Sunset	S. & W.	8	......	Clear.
Sept. 5	24 ...do...	35.17	30.95	1.5	8 a.m.	N. & E.	1	......	do.
Sept. 14	25 ...do...	34.62	31.17	10.6	Sunset	N. & E.	0	37.86	Clear at sunset; rained from 3 to 4 p.m.
Oct. 15	34 ...do...	34.69	31.21	1.9	6h. 45m. a.m.	N. & E.	1		Thick fog.
Oct. 4	38 same as 25	34.90	30.73	0.2	7 a.m.	N. & E.	1	......	Clear.
Sept. 15	26 ...do...	34.42	31.02	11.6	Sunset		0	40.19	Slightly smoky.
Sept. 16	26 ...do...	34.32	31.03	3	6 a.m.		0		Foggy and cloudy.
Oct. 2	37 same as 26	34.55	30.47	23.2	Sunrise	N. & E.	0	43.93	Clear.
Oct. 3	26 ...do...	34.52	30.70	6	Sunset	W.	1	......	Smoky.
Sept. 16	27 ...do...	34.33	30.93	12.3	Sunrise	N. & E.	2	......	Slightly cloudy and smoky.
Oct. 17	27 ...do...	34.37	31.06	6	...do...		1	......	Smoky.
Oct. 18	28 Near Goose lake	34.45	30.78	3.7	...do...		0	44.97	do.
Oct. 19	28 ...do...	33.90	30.49	14.5	Sunrise	N. & E.	0	......	Slightly clear and smoky.
Oct. 19	29 ...do...	33.95	30.52	2.6	9h. 30m. a.m.		1	......	Smoky.
Oct. 19	29 ...do...	34	30.53	15.8	Noon	N. & E.	1	......	Slightly cloudy.
Oct. 19	29 ...do...	33.95	30.52	20.5	Sunset	N. & E.	0	......	do.
Oct. 20	29 ...do...	33.60	30.51	17.5	Sunrise		0	......	do.
Oct. 20	29 ...do...	34.51	30.51	3.6	Sunset		9	......	do.
Oct. 30	29 ...do...	33.46	30.46	17.9	Sunrise	N. & E.	1	......	do.
Oct. 30	29 ...do...	33.98	30.50	6.8	Noon	N. & E.	3	......	Clear and cloudy and smoky.
Oct. 30	29 ...do...	33.84	30.40	19.7	Sunset	N. & E.	1	......	do.
Oct. 1	23 ...do...	33.85	30.49	14.2	Sunrise	N. & E.	4	......	do.

Meteorological observations—Continued.

Date	No. of camp	Location of camp.	Barometer. Upper reading. Cmt.	Barometer. Lower reading. Cmt.	Thermometer. Attached.	Thermometer. Detached.	Time of observation.	Wind. Direction.	Wind. Intensity.	Altitude above Lassen's, in feet.	Remarks.
1849. Sept. 20	30	Near Goose lake	34.05	30.63	14.5	.....	Sunset	W.	6	42.95	Slightly cloudy and smoky.
21	30	Near Goose lake	34.00	30.75	3.2	.....	Sunrise	N. & W.	1	.....	Clear; no dew; smoky.
21	.....	Foot of the west slope of Sierra Nevada, near head of Goose lake	34.06	30.51	28.8	.....	9h. 20m. p. m.	N. & W.	Gentle Puffs	46.14	...do......do....
21	31	.....	33.84	30.35	30.7	.....	Sunset	W.	1	.....	Clear; slightly smoky.
22	31	One mile east of summit of Sierra Nevada	33.63	30.38	1.5	.....	Sunrise	W.	0	45.70	...do......do....
22	.....	.....	34.04	30.60	19	.....	8 a. m.	W.	1	.....	Clear; no dew.
22	.....	.....	34.55	31.05	24.5	.....	6h. 20m. p. m.	.....	0	44.60	Clear.
23	.....	.....	34.59	31.24	18	.....	Sunrise	N.	0	38.05	..do....
23	.....	.....	34.40	31.11	3.3	.....	Sunset	.....	1	.....	..do....
24	.....	.....	34.37	30.96	18.4	.....	Sunrise	S.	0	.....	Thin clouds.
24	.....	.....	34.29	31.05	0.9	.....	Sunset	W.	1	.....	Clear.
25	.....	.....	33.96	30.67	1.8	.....	...do.	.....	1	.....	..do....

Note.—In the column headed "intensity of wind" in the foregoing record, 0 signifies calm; 1 signifies barely perceptible breeze; 2 signifies gentle breeze; 3 signifies moderate breeze; 4 signifies brisk breeze; 5 signifies strong wind.

The barometer used gives the altitude of the mercurial column 0.53 centimeters, too great according to a comparison with a standard barometer in Paris. The calculations of altitudes of camps were made by me in camp with the best means I had at my command. They are presented as near approximations to the true results.

R. S. WILLIAMSON,
Lieutenant U. S. Topographical Engineers.

Brevet Lieut. Col. J. HOOKER,
Assistant Adjutant General, Pacific division.

Astronomical observations taken during Captain Warner's explorations in the Sierra Nevada, 1849.

CAMP 12.—DEER CREEK, AUGUST 24, 1849.

EQUAL ALTITUDES OF SUN'S UPPER LIMB.

Mean of 13 observations.

Time—morning.			Time—evening.		
h.	m.	s.	h.	m.	s.
8	37	44.9	3	00	44.7

Chronometer slow 8m. 30.4s.

CAMP 12.—DEER CREEK, AUGUST 24, 1849.

CIRCUMMERIDIAN ALTITUDES OF SUN'S UPPER LIMB.

	Mean of observed altitudes.			Chronometer time of observations.		
	Deg.	min.	sec.	h.	m.	s.
1				11	39	7.4
2				11	40	8.2
3				11	41	2.2
4				11	41	55.2
5				11	42	48.6
6				11	43	42.6
7	61	14	54.5	11	44	47.4
8				11	45	39.4
9				11	46	41
10				11	47	38.2
11				11	48	23.4
12				11	49	21
13				11	50	11.4
14				11	50	57.8
15				11	51	44.6
16				11	52	30.2
17				11	53	26.2
18				11	54	24.2
19				11	55	15
20				11	56	6.6
21				11	56	50.6
22				11	57	33.8
23				11	58	34.2
24				11	59	27
25				12	00	16.2

Barometer, 29.84 inches; thermometer, 96°; latitude camp 12, 39° 56′ 36.9″.

Astronomical observations taken during Captain Warner's explorations
the Sierra Nevada, 1849.

DEER CREEK, AUGUST 25, 1849.

EQUAL ALTITUDES OF SUN'S UPPER LIMB.

Mean of 13 observations.

Time—morning.			Time—evening.		
h.	min.	sec.	h.	min.	sec.
8	26	57	3	13	13.1

Chronometer slow 7 min. 55.9 sec.

DEER CREEK, AUGUST 25, 1849.

CIRCUMMERIDIAN ALTITUDES OF SUN'S UPPER LIMB.

	Mean of observed altitudes.			Chronometer time of observation.		
	Deg.	min.	sec.	h.	min.	sec.
1				11	42	25.4
2				11	43	3.8
3				11	43	33.8
4				11	44	19
5				11	44	57.4
6	60	54	37.6	11	45	29.4
7				11	45	58.2
8				11	46	43.4
9				11	47	27
10				11	48	5.8
11				11	48	45
12				11	49	31.8
13				11	49	55
14				11	50	32.6
15				11	51	18.2
16				11	52	14.6
17				11	53	7
18				11	53	37.4
19				11	54	28.2
20				11	55	25.4
21				11	56	6.6
22				11	57	8.6
23				11	57	51.4
24				11	58	36.2
25				11	59	44.2

Barometer, 29. 88 inches; thermometer, 94°; latitude camp 12, 39° 56' 16" N.

Astronomical observations taken during Captain Warner's explorations in the Sierra Nevada, 1849.

CAMP 12, DEER CREEK, AUGUST 26, 1849.

CIRCUMMERIDIAN ALTITUDES OF SUN'S UPPER LIMB.

	Mean of altitudes.			Chronometer time of observation.		
	Deg.	min.	sec.	h.	min.	sec.
1				11	45	33.4
2				11	46	49
3				11	46	40.6
4				11	47	15.4
5				11	47	50.2
6				11	48	25.6
7				11	49	17
8				11	50	1
9	60	34	25.5	11	50	39.8
10				11	51	21.8
11				11	51	59
12				11	52	41.8
13				11	53	33
14				11	54	7.2
15				11	54	49
16				11	55	40.2
17				11	56	26.6
18				11	57	6.6
19				11	57	39
20				11	58	14.2

Chronometer time of apparent noon, 11h. 51m. 7s.; barometer, 29.91 inches; thermometer, 95°; latitude of camp 12, 39° 56' 9.3" N. Mean of three calculated latitudes of camp 12, 39° 56' 27.4" N.

CAMP 29, NEAR GOOSE LAKE, SEPTEMBER 19, 1849.

EQUAL ALTITUDES OF SUN'S UPPER LIMB.

Mean of 13 observations.

Time—morning.			Time—evening.		
h.	min.	sec.	h.	min.	sec.
9	13	0.6	2	27	4.6

Chronometer slow 16m. 7.27s.

Astronomical observations taken during Captain Warner's explorations in the Sierra Nevada, 1849.

CAMP 29, NEAR GOOSE LAKE, SEPTEMBER 19, 1849.

CIRCUMMERIDIAN ALTITUDES OF SUN'S UPPER LIMB.

	Mean of observed altitudes.			Chronometer time of observation.		
	Deg.	min.	sec.	h.	min.	sec.
1				11	48	46.4
2				11	49	50
3				11	50	33.2
4				11	51	13.6
5				11	51	51.6
6				11	52	30
7				11	53	4.8
8	49	49	8.7	11	53	52
9				11	54	41.2
10				11	55	19.2
11				11	56	6.4
12				11	56	41.2
13				11	57	22
14				11	58	23.8
15				11	58	58.8

Barometer, 25.38 inches; thermometer, 69°; latitude camp 29, 41° 43' 34.6".

The preceding observations were taken by Captain Warner with a sextant made by Gambey. The timepiece used was a pocket chronometer, No. 739, of Brockbanks. We found our chronometers to vary so much, from being carried on mules, that we could not rely upon them for longitude.

R. S. WILLIAMSON,
Lieutenant Topographical Engineers.

Brevet Lieutenant Colonel J. HOOKER,
Assistant Adjutant General, Pacific Division.

Distances between points in the Sacramento valley, as measured with an odometer, on the usually-travelled roads.

From—	To—	Miles.	Remarks.
Benicia	Suisun rancho	13.91	
	Berry's	15.85	
	Vacas	24.60	
	Puta creek	43.51	Following the road which leads across the tule to Sacramento City.
	Sacramento City	60.14	
Sacramento City	Suterville	3.00	Not measured with odometer.
Suterville	Sutter's Fort	2.86	
	Crossing of American Fork at Childs's	5.40	
	Nicholas, on Feather river	31.54	
	Crossing of Bear creek	35.18	
	Yuba river	51.73	
	Crossing of Feather river	66.65	
	Bend of Feather river	74.12	
	Chase's, or Niell's, on Butte creek	90.36	
	Little Butte creek	93.62	
	Lassen's, on Deer creek	116.65	
Lassen's	Crossing of Sacramento	3.93	
	Stone creek	21.74	
	Williams's	45.00	
	Slough of Sacramento	65.43	Going in nearly a straight line, while the wagon road follows the turn of the river.
Rancho of Zone	Rancho of Lone Tree	82.54	
	Paddy Clark's, on Cash creek	92.29	
	Wolf's Kills, on Puta creek	108.80	Several miles above where the road crosses. Joined the road 5 miles from Paddy Clark's.
	Puta creek	....	
	Vacas	119.65	
	Benicia	144.24	

Distances on the Lassen's trail between Lassen's Rancho, on Deer creek, and the eastern spur of the Sierra Nevada.

From—	To—	Miles.	Total distance from Lassen's.	Remarks.
Lassen's	Base of the Sierra Nevada at Dry creek	6¼	6¼	Water is almost always found in Dry creek, though it sometimes sinks upon leaving the mountain.
Base of Sierra	Mill creek	4¾	11	The descent can be made at this point, and animals taken to water. The road follows a ridge between two precipitous ravines. In the right one flows Mill creek; in the left, Dry creek. The first water on the road is 38 miles from Lassen's; but there are two or three places where animals may be driven down to Dry creek, besides the point on Mill creek above mentioned. Road very rough, rocky, and hilly, and destitute of grass, and but very little to be found in ravines.
Mill creek	A point of practicable descent to Dry creek	9	20	
Dry creek	First water on road	18	38	This is a small run of water, about a mile beyond the summit of the western spur of the Sierra Nevada.
First water	Headwaters of Deer creek	12	50	Deer creek and Butte creek both rise in the middle of a narrow valley, 10 miles in length, and run in contrary directions at first. During the 12 miles last mentioned, there is water to be found every three or four miles.
Deer creek	South fork Feather river	17	67	Last 7 miles hilly and rocky; first 10 very good road.
South fork Feather river	North fork Feather river	11	78	Road pretty rough; water and grass plenty during first five miles.
North fork Feather river	Small lakes	19	97	A few miles before reaching these lakes, the road forks to the left, and leads to a valley through which runs a stream of water. The roads meet again before reaching the lakes.
Lakes	Round valley, with grass and water	13	110	Road nearly level, and pretty good.
Valley	Spring to left of road	14	124	
Spring	Small stream	6	130	
Stream	Another small stream	6	136	
Branch of Pit river	Pit river	12	148	Said to empty into Pit river. Last 12 miles of road good. This 12 miles is the worst portion of the whole road.

| | | 80 | 228 | Along Pit river the road is nearly level, and generally good; and a good camping place may be found every few miles. The Lassen trail crosses the spur of the Sierra near Goose lake, the mountains being 1,500 feet above the valley of Goose lake, and about 6 miles from base to base. Several small streams run from the mountains into Goose Lake valley; one of which sinks, and another joins with springs from which Pit river flows. Pit river rises within three miles of Goose lake, but seems to have no communication with that body of salt water. |
| Pit river...... | ..... Goose Lake.............. | | | |

R. S. WILLIAMSON,
Lieutenant United States Topographical Engineers.

Brevet Colonel J. HOOKER,
Assistant Adjutant General, Pacific Division.

Report of Lieutenant R. S. Williamson, Topographical Engineers.

HEADQUARTERS THIRD DIVISION,
Benicia, March 27, 1850.

The following report is respectfully forwarded.

PERSIFOR F. SMITH,
Brevet Major General Commanding.

—

BENICIA, CALIFORNIA,
March 23, 1850.

SIR: I have the honor to report, for the information of the major general commanding this division, the result of an exploration of Monte Diablo, and the valley lying between this mountain and the southern shore of Suisun bay. This mountain, from its isolated position, affords from the summit an extensive view on all sides. The towns of San José, San Francisco, Sonoma, Happa, Benicia, New York, Stockton, Sacramento City, Termon, are all visible from there on a clear day; and it was my desire to have taken the bearings of all these different towns, and also of the prominent headlands in the Sierra Nevada and coast range. A general examination of the geological structure of the mountain, observations for the determination of its altitude, and an examination of Monte Diablo creek—a small stream, which, rising in the mountain, empties into Suisun bay—entered also into my plans. Mr. Bomford, a young gentleman who has made considerable proficiency in the science of geology, was my companion upon the expedition, and attended principally to this branch of our examination.

We started from Benicia early in December, with four or five days' provisions; crossed the straits of Carquines in a small boat belonging to this post, and, entering Monte Diablo creek about the time the tide had fairly commenced to run up, we succeeded in getting our boat, containing eight persons, about six miles up the creek. This stream during the dry season receives no water from the mountain, and is then, in fact, a slough, which is nearly dry at low water; but, by taking advantage of the tide, it is navigable for four miles at all seasons for launches and other boats drawing but two and a half feet water. It is probable that, in the spring, the body of water flowing into it from the mountain and adjacent hills gives it sufficient depth to admit vessels higher up and of greater draught. The land on either bank of the creek, for some distance from its mouth, is a level bed of tulé, and it is only when the stream enters the slightly elevated ground beyond, that it commences to become shallow; so that at its head of navigation the banks are sufficiently elevated above the bay to be easily approached by wagons. We encamped that night on the bank of the creek, and the following morning, leaving with the boat a party of officers who had availed themselves of this opportunity to go on a shooting excursion, we went to a Californian's rancho, about two miles distant, to hire a couple of horses for the transportation of bedding, provisions, &c. This rancho, called the rancho of Monte Diablo, or Pachico's rancho, lies from four to five miles from the base of the mountain. So much time had been consumed in catching the horses, that it was 2 o'clock before we were enabled to make a fair start. Our path at first lay over a succession of hills, separated by ravines, and at sunset, instead of being near the summit, as we had anticipated, we found ourselves at a very small elevation above the valley. The next morning at daylight we were preparing for the ascent, which we found quite difficult. After toiling up a steep ascent, we were frequently disappointed at finding a precipitous ravine before us, into which we would have to descend only to ascend again on the other side, and the sun again set without our having quite reached the summit. We had, however, but three or four hundred feet more to ascend, and were very well pleased with our progress. We arose before daylight on the following morning, and, as the dawn approached, a sublime, but, to me, rather discouraging sight was presented. The whole of the surrounding valley was covered with a dense fog, the tops of the higher hills here and there breaking through. As the sun arose, a new beauty was added to the scene. We were in hopes that the fog would clear away, under the influence of the sun's rays, but this expectation was but partially realized. The fog continued to hang over the whole country to the westward, though the valleys to the eastward were spread out to our view. We walked to the summit and found the very highest point easily accessible. nd a d an *unobstructed* view in every direction. I took the bearings of the prominent points not hidden by the fog, and took, also, several barometric observations during the course of the morning, which, with those taken near the base of the mountain, both previously and subsequently, gave me the altitude, 3,960 feet. As our provisions and water were both exhausted, we were unable to remain, in hopes of having a clear day on the morrow, but reluctantly commenced the descent, and the following day arrived at Benicia.

During the whole of this exploration, Mr. Bomford was indefatigable in searching for mineralogical specimens, and made a discovery which may hereafter be very valuable. Upon almost every part of the mountain he found specimens of calcareous stones. Near the very summit some beautiful ones were collected. They were small irregular masses of pure white crystalline stone, imbedded in the body of the rock comprising the principal mass of the mountain top. Very near the base of the mountain was found the largest deposite of calcareous stone that we saw; and from this place several specimens were brought away to experiment upon. The rock from which these specimens were taken was 20 feet high, and appeared to be the outcrop of a large ridge. After our return to Benicia, Mr. Bomford and myself both experimented upon the specimens from this rock, each having in view to ascertain if the stone contained impurities, and the nature of them; but proceeding to this result by different means. Mr. Bomford took several pieces of the stone, each about the size of a walnut, placed them in a crucible, and subjected them to a strong regular heat in a blacksmith's forge—some for four, some for five, and some for six hours. He produced, in either case, a lime which slaked freely by immersion or sprinkling, forming a thick paste, which was generally smooth to the feel; all of which tended to show that the stone was a nearly pure limestone, and consequently possessed no hydraulic properties. In my experiments, performed at the same time, I endeavored to ascertain by analysis the quantity of those impurities of the stone, if any existed, from which an inference might be drawn as regards its hydraulic properties. To effect this, I pulverized a portion of the stone and weighed carefully five drachms of the powder, which I placed in a small glass mortar and treated to muriatic acid. The solution advanced rapidly and with much effervescence, and I soon had a clear liquid, which appeared to have a very slight sediment only. This liquid was then partially evaporated in a sand-bath, and, the residue having been mixed with a pint of pure water, was then filtered. The residuum left on the filtering paper was evidently all the matter in the five drachms of limestone that had not been dissolved by the muriatic acid. Among this residuum, therefore, was all the allumina and silica in the stone; and these are the main agents in rendering a lime hydraulic. The amount of the residuum was so small that I could not collect and weigh it; but I am sure it was much less than one grain. This proved conclusively that the stone was almost entirely destitute of earthy and silicious matter. Another active agent in producing hydraulic action is magnesia; and had any existed in the stone, it would have been dissolved with the lime, and would not have been present in the residuum on the filtering paper. It was necessary, then, to ascertain, by a chemical test applied to the filtered solution, if any magnesia was present. Muriatic acid, as above stated, dissolves both lime and magnesia, but the acid has a much greater affinity to the former; and if, in a solution of magnesia in this acid, a quantity of lime-water be added, the chlorine of the acid, before combined with the magnesium, will leave this base and attach itself to the calcium of the lime, and cause a precipitate of magnesia. A strong lime-water was made, and a small port on of the filtered solution tested with it. Had any magnesia existed, it must, according to the above-mentioned principle, have been precipitated; but no precipitate was found, and the liquid remained clear. Had iron existed in the solution, it would also have been precipitated by the lime-water. The inference drawn from these experiments is, that neither allumina, silica, magnesia, or iron, exists in appreciable quantities in this stone; that it is nearly pure carbonate of lime, and yields a fat lime of excellent quality for the ordinary purposes of construction, but possesses no hydraulic properties. I regret that the relative weight of the stone before and after calcination was not observed; but think the stone must have lost very nearly half its weight in burning. The mass from which the specimens tested were taken is elevated between 500 and 600 feet above Monte Diablo valley, and is distant about six miles from the head of navigation of the creek. There is timber in the vicinity, though not in great abundance, being oaks scattered over the hills and valley, as is common in this part of the country.

In connexion with this subject, I respectfully report that I visited to-day the lime kiln established about four miles from this place on the Suisun road, and examined the stone, the manner in which it existed, and the lime it produced. It is found in sedimentary deposites, lying imbedded in sandstone. It is not found below three feet from the surface, and the deposites are quite small. The stone itself is generally impure, as is apparent to the eye, containing considerable quantities of silicious matter.

A favorable specimen of the calcined stone, taken from the kiln just over the furnace, would not slake at all. This had undoubtedly been too much exposed to heat; but as pure limestone is not injured by being overburnt, this was another proof of the impurity of this stone. From the small quantity and impure quality of this stone, I am satisfied that the deposite on

CPSIA information can be obtained
at www.ICGtesting.com
Printed in the USA
BVHW04*0743011018
528775BV00026B/181/P